Proceedings of Wivace 2008

Artificial Life and Evolutionary Computation

Proceedings of Wivace 2008

Artificial Life and Evolutionary Computation

Venice, Italy   8 – 10 September 2008

Editors

Roberto Serra
Marco Villani
University of Modena and Reggio Emilia, Italy

Irene Poli
Ca' Foscari University, Italy

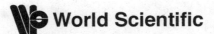 World Scientific

NEW JERSEY · LONDON · SINGAPORE · BEIJING · SHANGHAI · HONG KONG · TAIPEI · CHENNAI

Published by

World Scientific Publishing Co. Pte. Ltd.

5 Toh Tuck Link, Singapore 596224

USA office: 27 Warren Street, Suite 401-402, Hackensack, NJ 07601

UK office: 57 Shelton Street, Covent Garden, London WC2H 9HE

British Library Cataloguing-in-Publication Data
A catalogue record for this book is available from the British Library.

ARTIFICIAL LIFE AND EVOLUTIONARY COMPUTATION
Proceedings of Wivace 2008

ISBN-13 978-981-4287-44-9
ISBN-10 981-4287-44-X

Printed in Singapore.

PREFACE

The Italian community interested in artificial life and the one active in evolutionary computation have decided a couple of years ago to merge their annual meetings in a single event, since there were many interesting overlaps concerning both topics and persons. This decision has proven fruitful, and there have been successful meetings in Siena and Scicli, before this year's one in Venice. Previously, separated workshops had been organized for artificial life in Cosenza and Rome, and for evolutionary computation in Milan.

The growth, both in quantity and quality, of the activities in these fields are the reasons why in the 2008 meeting we decided to move to a more international setting, with some invited speakers and some participants coming from abroad, and to publish a proceedings volume.

We wish to thank all the invited speakers (Armando Bazzani, Annamaria Colacci, Rosaria Conte, Stuart Kauffman and Norman Packard) for their contribution to the success of the conference: besides delivering extremely interesting presentations, they all actively participated to the discussions in the conference room and also had many informal talks with the participants during coffee break and spare time.

This volume comprises all the accepted full-length papers which have been contributed to the Wivace08 workshop held in Venice, 8-10 September 2008, and also three invited papers. The workshop has been really successful, there have been many interesting discussions both in the conference room and in informal gatherings, and we hope that some new collaboration will spring from the encounters among the researchers which were in Venice. The Italian word "vivace" means lively, and the workshop Wivace stood up to its name!

We wish to express our gratitude to the organizations which made this event possible, the European Centre for Living Technology (ECLT) of the Ca' Foscari University in Venice and the Department of Social, Cognitive and Quantitative Sciences of the University of Modena and Reggio Emilia. The energy and competence of the ECLT operations manager, Elena Lynch, was particularly important in getting an effective organization and a smooth running of the workshop.

We are also grateful to the E-learning Centre the University of Modena and Reggio Emilia, and to its director Tommaso Minerva, for support in the preparation of the conference website.

Special thanks are due to the persons who most contributed to the organization of the workshop, our Ph.D. students Alex Graudenzi and Chiara Damiani, who took care with enthusiasm and skill of the various aspects related to the workshop and who collected all the contributions for this volume. Important support was also provided by Alessandro Filisetti and Luca Ansaloni.

We also wish to express our gratitude to the administrative staff of ECLT, in particular to Tatiana Zottino for a proper handling of all the financial aspects.

Finally, we thank Chelsea Chin at World Scientific for her friendly and careful assistance in the preparation of this volume.

Roberto Serra[a], Marco Villani[b], Irene Poli[c]

[a,b] *Department of Social, Cognitive and Quantitative Sciences, University of Modena and Reggio Emilia, via Allegri 9, 42100, Reggio Emilia, Italy. E-mail: rserra@unimore.it, mvillani@unimore.it*
[c] *Department of statistics, University Ca' Foscari of Venice, Cannaregio 873, 30121 Venice Italy; European Centre for Living Technology, San Marco 2940, 30124 Venice, Italy*

CONTENTS

Organizational committee

Roberto Serra - Università di Modena e Reggio Emilia (*chair*)

Irene Poli - Università Ca' Foscari di Venezia (*co-chair*)

Stefano Cagnoni - Università di Parma

Leonardo Vanneschi - Università di Milano Bicocca

Marco Villani - Università di Modena e Reggio Emilia

Promotional committee

Gianluca Baldassarre - Istituto di Scienze e Tecnologie della Cognizione - CNR

Eleonora Bilotta - Università della Calabria

Stefano Cagnoni - Università di Parma

Davide Marocco - Istituto di Scienze e Tecnologie della Cognizione - CNR

Marco Mirolli - Università di Siena; Istituto di Scienze e Tecnologie della Cognizione - CNR

Giuseppe Nicosia - Università di Catania

Stefano Nolfi - Istituto di Scienze e Tecnologie della Cognizione - CNR

Pietro Pantano - Università della Calabria

Domenico Parisi - Istituto di Scienze e Tecnologie della Cognizione - CNR

Tommaso Minerva - Università di Modena e Reggio

Program committee

Alberto Acerbi - Università di Siena, Istituto di Scienze e Tecnologie della Cognizione - CNR

Mauro Annunziato - Plancton Art Studio; Supervision and Advanced Control Lab - ENEA

Paolo Arena - Università di Catania

Antonia Azzini - Università di Milano

Gianluca Baldassarre - Istituto di Scienze e Tecnologie della Cognizione - CNR

Eleonora Bilotta - Università della Calabria

Andrea Bonarini - Politecnico di Milano

Anna Borghi - Università di Bologna

Ernesto Burattini - Università di Napoli "Federico II"

Stefano Cagnoni - Università di Parma

Raffaele Calabretta - Istituto di Scienze e Tecnologie della Cognizione - CNR

Angelo Cangelosi - University of Plymouth

Riccardo Caponetto - Università di Catania

Timoteo Carletti - Università di Namur (Belgio)

Maurizio Cardaci - Università di Palermo

Mauro Ceruti - Università di Bergamo

Antonio Chella - Università di Palermo

Roberto Cordeschi - Università di Roma "La Sapienza"

Vincenzo Cutello - Università di Catania

Ivano De Falco - Istituto di Calcolo e Reti ad Alte Prestazioni - CNR

Antonio Della Cioppa - Università di Salerno

Cecilia Di Chio - University of Essex

Michele Di Francesco - Università "Vita-Salute S.Raffaele" Milano

Salvatore Di Gregorio - Università della Calabria

Luigi Fadiga - Università di Parma

Maurizo Ferraris - Università di Torino

Mauro Francaviglia - Università di Torino

Gianluigi Folino - Istituto di Calcolo e Reti ad Alte Prestazioni (ICAR) - CNR

Francesco Fontanella - Università di Cassino

Luigi Fortuna - Università di Catania

Mauro Gallegati - Università di Ancona

Mario Giacobini - Università di Torino

Simone Giansante - CSC - Università di Siena; University of Essex

Stefano Ghirlanda - Università di Bologna

Alessandro Londei - Istituto di Scienze e Tecnologie della Cognizione - CNR

Davide Marocco - Istituto di Scienze e Tecnologie della Cognizione - CNR

Filippo Menolascina - National Cancer Institute "Giovanni Paolo II", Bari

Orazio Miglino - Università di Napoli "Federico II"

Tommaso Minerva - Università di Modena e Reggio

Marco Mirolli - Università di Siena; Istituto di Scienze e Tecnologie della Cognizione - CNR

Sandro Nannini - Università di Siena

Giuseppe Narzisi - NYU, USA

Giuseppe Nicosia - Università di Catania

Stefano Nolfi - Istituto di Scienze e Tecnologie della Cognizione - CNR

Peter Oliveto - University of Birmigham, UK

Enrico Pagello - Università di Padova

Luigi Pagliarini - Maersk Mc-Kinney Moller Institute; Accademia di Belle Arti di Bari

Pietro Pantano - Università della Calabria

Domenico Parisi - Istituto di Scienze e Tecnologie della Cognizione - CNR

Mario Pavone - Università di Catania

Alessio Plebe - Università di Messina

Irene Poli - Università Ca' Foscari di Venezia

Rosario Rascunà - University of Sussex, UK

Giuseppe Scollo - Università di Catania

Roberto Serra - Università di Modena e Reggio Emilia

Celestino Soddu - Politecnico di Milano

Serena Sordi - Università di Siena

Giandomenico Spezzano - Istituto di Calcolo e Reti ad Alte Prestazioni (ICAR) - CNR

Guglielmo Tamburrini - Università di Napoli "Federico II"

Pietro Terna - Univeristà di Torino

Andrea G.B. Tettamanzi - Dipartimento di Tecnologie dell'Informazione - Università degli Studi di Milano

Giorgio Turchetti - Università di Bologna

Leonardo Vanneschi - Università di Milano Bicocca

Alessandro Vercelli - Università di Siena

Marco Villani (Università di Modena e Reggio Emilia)

Luca Zammataro - Institute for Cancer Research and Treatment - Torino

Marco Zorzi - Università di Padova

Local operative committee

Luca Ansaloni (Università di Modena e Reggio Emilia)

Chiara Damiani (Università di Modena e Reggio Emilia)

Alessandro Filisetti (European Centre for Living Technology)

Alex Graudenzi (Università di Modena e Reggio Emilia)

Elena Lynch (European Centre for Living Technology)

Irene Poli (Università Ca' Foscari di Venezia)

Roberto Serra (Università di Modena e Reggio Emilia)

Marco Villani (Università di Modena e Reggio Emilia)

PART I

INVITED PAPERS

COGNITIVE DYNAMICS IN AN AUTOMATA GAS

A. BAZZANI*, B. GIORGINI, F. ZANLUNGO and S. RAMBALDI

University of Bologna - L.Galvani Center for Biocomplexity - Laboratory "Fisica della Città"
and INFN Bologna
Physics Department, Via Irnerio 46, 40126 Bologna, Italy
* E-mail: armando.bazzani@bo.infn.it
www.physicsofthecitylab.unibo.it

The problem of modeling cognitive systems has been recently considered under new point of views. The attention is no more mainly focused on the simulation of rational behaviors, but also on the importance of the individual free-will and of the apparent "irrational decisions". A possible approach for a mathematical modeling is to introduce an internal dynamical structure for the "elementary particles" that allows to define an individual cognitive dynamics at the base of the decision mechanisms. Our assumptions imply the existence of an utility potential which is the result of an evolution process, where some behavioral strategies have been selected to perform efficiently in presence of different environmental conditions. The utility potential defines a cognitive dynamics as a consequence of information based interactions and the choice of a strategy depends on the individual internal cognitive state. This is the definition of an "automata gas model", whose aim is to extend some statistical physics results to cognitive systems. We shall discuss the existence of self-organized dynamical states in an automata gas, that are the result of a cognitive behavior. We show as the introduction of a cognitive internal dynamics allows to simulate the individual rationality and implies the existence of critical conditions at which the system performs macroscopic phase transitions. The application to the pedestrian mobility modeling problem is explicitly considered and we discuss the possibility of a comparison with experimental observations.

Keywords: Cognitive Behavior; Automata Gas; Pedestrians Dynamics.

1. Introduction

Any simple decision is the result of complex processes that takes into account the available information, the previous experience and the individual propensities. The rational decision models, that are mainly derived from the game theory,[22] are not suitable to simulate the individual free-will and the existence of different strategies in each individual.[16] These phenomena could play an essential role to explain both the adaptation capability of cognitive systems to dynamical environments, and the existence of sudden transitions in the macroscopic dynamical states, like the develop of panic in crowded or critical conditions.[13] A possible approach to a mathematical modeling of cognitive systems may take advantage from the idea of *cognitive dynamics* recently proposed by neuroscientists.[7] This theory assumes the existence of an internal cognitive state, which creates an interface between the external stimuli and the consequent decisions and whose evolution simulates the brain activity. On one hand the introduction of a cognitive state is quite natural if one interprets the brain activity as a dynamical system, coupled with the external environment by information based interactions.[24] On the other hand the idea of propensity understands the existence of an utility function, that allows a quantification of the expected utility of a certain choice taking into account the available information and the previous experience.[5] We consider the utility function as the result of an evolution process and as a common feature to all the individuals. Any rational individual will choice the decision that maximized the estimated utility, but due to the existence of an individual cognitive state, it is not possible to define a universal rationality, since the state depends on the previous decisions and the information at disposal. Using a parallel with the concept of *Subjective Probability* introduced by B. deFinetti,[4] we can say that only a subjective rationality exists.

In the simple examples we consider, a population of automata has to decide between two possible choices.[2] Each automaton changes its cognitive state according to a Langevin's equation defined by an utility potential which depends on the external information.[25] A stochastic noise reproduces the effect of the continuous unpredictable interactions not directly related to the decision utility, but that can modify the cognitive state. The noise amplitude can be also interpreted as the individual attitude to change mind. Each cognitive state is associated to the individual propensity toward one of the possible choices and defines the decision mechanism in a probabilistic way. The decision realization means to perform a particular action, which allows to get new information and, consequently, to change the utility potential. In this way the automata perform a learning procedure that enforces the most successful choice and creates self-organized dynamical states due to coupling between physical and cognitive dynamics.[26,27]

The aim of this paper is to study some statistical properties and the presence of different strategies in an automata gas inspired by the pedestrian dynamics modeling. Such a problem has attracted the attention of physicists and mathematicians since several years ago,[3,8] due to its relevance both for safety reasons (risk control during crowd stampede and panic situations) and for improving the accessibility of urban space like stations, museums or shopping centers. In the second section we discuss a possible mathematical modeling of an automata gas for pedestrian mobility in simple urban space. The third section we apply a statistical physics approach to study the existence of critical values in the system parameters for the appearance of emergent states and we illustrate some properties of the model by using numerical simulations. Finally in the last section we discuss the problems related to experimental validations of the automata gas model by using video data recorded during the Venezia Carnival 2007.[15]

2. The Automata Gas Model

The automata gas model simulates an ensemble of cognitive particles moving in a given space. Using a reductionist point of view, we divided the interactions in two main classes: the physical and the cognitive interactions. By physical interactions we mean not only the effect of physical forces, but also the dynamical phenomena that can be described by Newton-like equations by assuming the existence of *social forces*.[10] The social forces simulate the effect automatic strategies, that are a direct response to external stimuli without the necessity of cognitive processes: a typical example is the collision avoidance strategy of pedestrians where the vision mechanism may be simulated by a long range repulsive force. In our model the automata have a incompressible body of dimension r_b, a social space of dimension $2r_b$ and a visual radius of dimension $\simeq 5r_b$ (fig. 1), that define the spatial scales of physical interactions. The social space is used to simulate the pedestrians attitude to avoid the physical contacts with other individuals, whereas the visual radius defines semicircle centered at the automaton velocity direction which reproduces the effect is of a frontal vision. Each automaton i tends to move at a desired velocity $\vec{v}_{0i}$ according to

$$\dot{\vec{v}}_i = -\gamma(\vec{v}_i - \vec{v}_{0i}) \tag{1}$$

where the parameter $1/\gamma$ defines the microscopic relaxation time scale of the system. We also provide the automata of an inertial mass m and we introduce inelastic collisions as a consequence of the automata body incompressibility.

The vision mechanism is realized by a topological long range interaction among the automata: each automaton focuses his attention to the other automata in the visual space, whose trajectories will enter its social space.

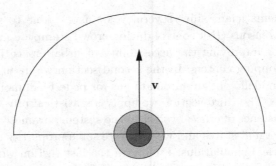

Fig. 1. Sketch of automaton physical dimensions: the dark circle is the incompressible body of radius r_b; the light circle is the social space of radius $2r_b$ and the larger semicircle is the frontal visual space of radius $5r_b$.

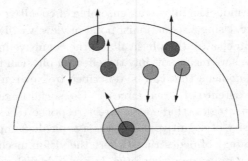

Fig. 2. Collision avoiding strategy of automata as a consequence of a local vision mechanism.

In such a case it rotates and/or reduces the velocity in order to avoid future collisions. The net result is a repulsive force among the automata coming from opposite directions, which does not satisfied the Action-Reaction principle (frontal vision). The repulsive force turns out to be proportional to the density of counteracting automata in the visual space and to the rotational velocity and deceleration rate values. Due to the long range character of local vision, the automaton has the tendency to move towards empty regions or to follow other automata moving in the same direction. This mechanism allows the formation of self-organized dynamical states: let us consider two counteracting automata flows along a narrow corridor. If the relaxation time $1/\gamma$ is not too small (in the limit $\gamma \to \infty$ the velocity is frozen along the desired direction), the collisions dynamics tends to distribute the automata population along the corridor according to maximal entropy principle (fig.

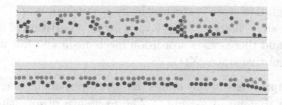

Fig. 3. Dynamics of two counteracting flow of automata moving along a corridor. When the automata interact only by means of inelastic collisions a disordered state is observed along the corridor(top); if the local vision is introduced, a two streams ordered flow appear along the corridor (bottom).

3 (top)). When the local vision is introduced, each automaton moves to avoid collisions and the system tends to relax to a stable dynamical state which minimizes the collisions number. As a result two ordered counteracting streams appear along corridor (fig. 3 (bottom)). The stability of the self-organized solution is destroyed when the automata density along the corridor overcomes a critical threshold, simulating the appearance of a panic dynamics, but this is a consequence of physical interactions as the transition from a laminar to a turbulent regime in fluids.

In order to introduce a cognitive behavior in the automata that cannot be reduced to a Newton's like dynamics we are inspired by recent experimental results in neuroscience research,[18] that can be summarized by the following remarks:

- the decisions are always mediated by the brain activity (existence of a cognitive space);
- there is a nonlinear subjective relation between the "utility" of a decision and the probability of taking that decision: an overestimate of small advantages and an underestimate of the large advantages with respect to a linear relation;
- there is an aversion to risk.

Then we associate to each automaton a dynamical cognitive model, that is formulated according the assumptions:

- the brain activity can be represented by a dynamical system defined on a n-dimensional cognitive space;
- the utility associated to a decision introduces a landscape potential in the cognitive space whose shape depends on the external information;

- the cognitive state is stochastically perturbed;
- the possible choices E_j $j = 1, ..., n$ define a partition of the cognitive space and there exists a decision mechanism which is a function of the cognitive states $X(t)$.

The first assumption essentially states that the brain can be macroscopically described by a dynamical system.[24] The existence of an utility function in the decision mechanism has been proposed in various contexts.[5] Our point of view is to consider the utility function as a potential $V(X; I)$ in the cognitive space, whose minima are related to the perceived utility of the various decisions depending on the information I. Of course there is a strict relation between self-organization and the "information" inserted in the utility function. Indeed the definition of "interesting information" for a certain decision is a key point to model a cognitive behavior.[25]

The third assumption is quite obvious since any individual decision can be influenced by several unpredictable factors.

Finally the last assumption allows to introduce a subjective rationality since the choice probabilities depend on the individual cognitive state X and/or information level. According to hypotheses, the automaton cognitive dynamics is defined by a stochastic differential equation

$$dX = -\frac{\partial V}{\partial X}(X; I)dt + \sqrt{2T}dw(t) \qquad (2)$$

where we have introduced the individual "social temperature" T (i.e. a measure of the individual influence of external random factors on the cognitive state) and $w(t)$ is a Wiener processes. The cognitive dynamics is the result of a deterministic tendency to increase the utility proportional to the "force" $\partial V/\partial X$, and of random effects to describe the great number of unpredictable factors that enter in any individual decision. When T is small, the stochastic dynamics $X(t)$ is essentially confined in a neighborhood of the potential minima and it is straightforward to associate each minimum to a particular choice and the well deepness to the choice utility. In other words, the cognitive space represents different brain areas and the utility defines the probability that a certain area is activated. The solution $X(t)$ is the evolution of the brain activity and the automaton decision will be associated to the choice E_i if

$$\frac{1}{\tau_d} \int_t^{t+\tau_d} \chi_{E_i}(X(s))\, ds = \max_{j=1,..,n} \left[\frac{1}{\tau_d} \int_t^{t+\tau_d} \chi_{E_j}(X(s))\, ds \right] \qquad (3)$$

where τ_d defines the decision time-scale and $\chi_{E_j}(X)$ is the characteristic function of the set E_j: i.e. the decision is associated to the most visited choice during the decision time. The information dependence of the potential $V(X; I)$ simulates a common rationality in the population, that modifies the utility function when any individual get certain information I. We can distinguish three different time scales in the model: the relaxation time scale (or Kramer's time scale) τ_K of the stochastic dynamics (2), the decision time scale τ_d and the information acquisition time scale. The relaxation time scale is the time requires to the dynamics (3) to reach a stationary state. The decision time can be interpreted as a reasoning time: if $\tau_d \ll \tau_K$ then the choice depends strongly on the actual cognitive state (subjective rationality), otherwise each automaton tends to select the most useful choice among all the possibilities (objective rationality). Finally the information acquisition time scale is directly related to the dynamical properties of the automata gas, since an automaton get new information as a consequence of a physical action. In the model we assume that τ_d is shorter than the information acquisition time scale, so that a long decision time scale corresponds to a delay in the information acquisition.

We will illustrate the properties of the definitions (2,3) in the atomic decision case (the decision between two possible choices). Then the utility potential reduces to a double well potential

$$V(X, I) = X^2 \left(\frac{X^2}{4} + \frac{X}{3}(X_A - X_B) - \frac{X_A X_B}{2} \right) \qquad X_{A,B} > 0 \qquad (4)$$

where the stable fixed points $-X_A(I), X_B(I)$ represent the two possible choices. The cognitive space partition is defined by $E_A = \{x \, / \, x < 0\}$ and $E_B = \{x \, / \, x > 0\}$. For given values X_A and X_B, we have considered an automata population in a stationary cognitive state. In the figure 4 we plot the probability of choosing B as a function of the utility $-V(X_B)$, for $X_A = 1$ constant. The circles refer to a short decision time, whereas the diamonds refer to a long decision time in. We remark as in the first case the empirical probability is quite sensitive to the utility changes when the utility is small and tends to saturate to 1 in a smooth way when the utility increases; on the contrary in the second case we see a threshold effect, since the population changes sharply from one choice to the other as the utility changes. According to eq. (4), the utility potential is a function of the relevant macroscopic information. We are interested in the case when two opposite strategies are present in the population: a cooperative strategy that tends to increase gradually the utility if the majority of the population tends to cooperate and a selfish strategy that suddenly decreases the utility if too many people take the same decision. The relevant information is the population fraction $P_{A,B} = N_{AB}/N$ that takes the same choice A or B

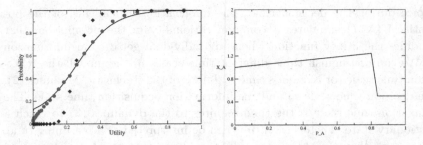

Fig. 4. Left figure: choice probability in the case of atomic decision as a function of the utility $-V(X_B)$; we have considered a population of 10^4 independent automata performing the cognitive dynamics (2) where $T = .1$ and the utility potential is defined by (4) ($X_A = 1$ constant). The circles refer to a short decision time (2.5 arbitrary times units), whereas the diamonds refer to a long decision time 250 time units; the continuous curve is the mean field estimate using the Kramer's transition rate theory (see eq. (9)). Right figure: dependence of the utility potential fixed point on the cooperative population fraction; the increasing behavior is related to a cooperative strategy, whereas the fast decrease introduces a selfish strategy in the population. The parameters are $P_* = .8$ and $a = 20$ (cfr. eq. (5)).

and we introduce the strategies in the model by varying the fixed point positions according to

$$X_A(P_A) = \begin{cases} X_A^0(1 + cP_A) & \text{if} \quad P_A \leq P_* \\ \max\left(X_A^0(1 + cP_* - a(P_A - P_*)), X_m\right) \end{cases} \tag{5}$$

where the parameter c defines the cooperation degree and a the strength of the selfish strategy (fig. 4 (right)). P_* and X_m are respectively the congestion threshold and a minimal value for the fixed point position (an analogous definition holds for X_B). The different slopes in fig. 4 reflect also the different character of the two strategies: the cooperative strategy is a behavioral strategy that has been selected since it is advantageous in many situations, whereas the selfish strategy is an individual safety strategy that an automaton applies in possible dangerous situations. Under this point of view the automata are quite sensitive to danger due to crowding. In the application to pedestrian dynamics, such a model would like to describe the cooperative phenomena observed in experimental data,[11] like herding effects or streams in crowd dynamics, but also the possibility of appearance of critical states in particular congested situations.

3. Statistical Properties

In order to study the statistical properties of the automata gas model (2,3) we consider the evolution of the probability density $\varrho_i(X,t)$ for each automaton i in its cognitive space. Then we define the global distribution function

$$\varrho(X,t) = \frac{1}{N} \sum_{i=1}^{N} \varrho_i(X,t) \tag{6}$$

where the sum runs over the automata population. Assuming that the information I in the utility potential (4) is functional of the distribution $\varrho(X,t)$, the cognitive dynamics decouples from the physical dynamics (of course this is a simplification to allow an analytical approach). The information $I = I(\varrho(X,t))$ introduces a coupling among the automata and it is possible to justify a system of nonlinear Fokker-Planck equations for the distributions ϱ_i under the assumption of a generalized law of large numbers[21]

$$\frac{\partial \varrho_i}{\partial t} = \frac{\partial}{\partial X}\frac{\partial V}{\partial X}(x, I(\varrho))\varrho_i + T_i \frac{\partial^2 \varrho_i}{\partial X^2} \qquad i = 1, ..., N \tag{7}$$

Using the double well potential (4), we get the self-consistent stationary solution of the system (7)

$$\varrho(X) = \frac{1}{N} \sum_{i=1}^{N} K_i \exp -\frac{V(X, I(\varrho))}{T_i} \tag{8}$$

where K_i are normalization constants. For each automaton the transition rate between the regions $E_A = \{x \, / \, x < 0\}$ and $E_B = \{x \, / \, x > 0\}$ associated to the possible choices is estimated by the Kramer's theory[9]

$$\pi_{AB} = \frac{|\omega_A \omega_0|}{2\pi} e^{V_A/T_i} \qquad \pi_{BA} = \frac{|\omega_B \omega_0|}{2\pi} e^{V_B/T_i} \tag{9}$$

where $\omega_{A,B,0}$ are the fixed point eigenvalues of the vector field $-\partial V/\partial X$. π_{AB}^{-1} and π_{BA}^{-1} define respectively the average time spent in the regions E_A and E_B (the Kramer's time scale τ_K). When the decision time τ_d is small compare, we can approximate the population fraction P_A according to

$$P_A = \int_{-\infty}^{0} \varrho(X) \, dX \tag{10}$$

This assumption can be numerically verified as it is shown in fig. 4 (left) where the formula (10) has been used to get the continuous curve. On the contrary when $\tau_d > \tau_K$, the most useful choice becomes more and more favored according to the decision mechanism (3); in the limit $\tau_d \to \infty$ all the automata will take the same decision according to

$$P_A = \begin{cases} 1 & \text{if } \int_{-\infty}^0 \varrho(X)\, dX > \frac{1}{2} \\ 0 & \text{otherwise} \end{cases} \tag{11}$$

We explicitly study the dependence (5) in the utility potential, assuming that the automata have the same social temperature $T_i = T$. Then from the solution (8), we get the self-consistent equation for P_A

$$P_A = K \int_{-\infty}^0 \exp\left(-\frac{V(x; P_A, P_B)}{T}\right) dx \qquad P_B = 1 - P_A \tag{12}$$

If we set $X_A^0 = X_B^0 = X^0$ (i.e. there is no a priori preferred choice), the trivial solution $P_A = P_B = 1/2$ always exists, but other solutions are possible. An explicit calculation can be performed by using the transition probabilities (9)

$$P_A = \frac{1}{1 + \dfrac{\omega_A}{\omega_B} \exp\left(\dfrac{V_A - V_B}{T}\right)} \tag{13}$$

where

$$\frac{\omega_A}{\omega_B} = \sqrt{\frac{X_A(P_A)}{X_B(P_B)}} \tag{14}$$

and $V_{A,B}$ are the utilities of the two choices. The equation can be solved by using a fixed point principle and a bifurcation phenomenon is observed at $c = c_*(T)$, when the following condition holds

$$\frac{\partial}{\partial n_A}\bigg|_{P_A=1/2} \frac{\omega_A}{\omega_C} \exp\left(\frac{V_A - V_C}{T}\right) = 4 \tag{15}$$

In the fig. 5, we plot P_A in the stationary state, as a function of the co-operation degree c, comparing Monte Carlo numerical simulations together with the analytical curve obtained from eq. (13). The existence of a bifurcation phenomenon means that a cooperative state emerges in the population only if the cooperative degree is sufficiently big for a fixed value of the social temperature. We have the limit $T \to 0$ as $c_* \to 0$, so that the automata tends to cooperate when the social temperature is small, but the time requires to relaxed to stationary state becomes extremely long according to the Kramer's estimate (9). These results are true for a small decision time τ_d, whereas the τ_d increasing causes a sharp transition to a high cooperative state. As a consequence the congested situations are also more probable, so

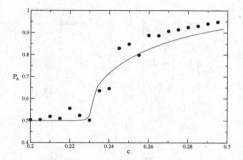

Fig. 5. Left figure: stationary value of the population fraction P_A in the case of atomic decision as a function cooperation degree c; we have considered the same parameters as in fig. 4 The circles refer to a short decision time (2.5 times units), whereas the continues curve is obtained from the Kramer's approach (13).

that we can observe the effects of the selfish strategy. This is illustrated by fig. 6 (right) where we plot the evolution of the P_A for different cooperation degrees when the decision time is long (250 time units). We observe that the transition to cooperative states emerges also for $c \leq c_*$, but as c increases, the appearance of congested states causes oscillations in the P_A values, since some automata feel the convenience of changing their decision.[28] The oscillations increase their amplitude according to the cooperative degree up to transition to a chaotic regime in the P_A dynamics (fig. 6), where the population majority changes her decision at random times. Moreover we remark the critical dependence of the cognitive dynamics on the c value. This behavior corresponds to a frustrated dynamics, in which the population is not able to relax to stationary state. When the cognitive dynamics is coupled with the physical dynamics the chaotic behavior may give rise to macroscopic transitions of the systems toward new stationary equilibria, but also to critical congested states which can decrease the system efficiency and create dangerous situations (for example the flux reduction across a bottleneck in pedestrian dynamics[14]) A semi-quantitative explanation of the numerical simulations plotted can be achieved by using a discrete version balance equation for the P_A fraction

$$P_A(t + \tau_d) = P_A(t) - \pi_{AB}P(t)\tau_d + \pi_{BA}(1 - P_A(t))\tau_d \qquad (16)$$

where π_{AB} and π_{BA} are the transition probabilities (9). In the figure 6 (right), we show the P_A evolution for different c values, using the same parameters as in the numerical simulations. We observe as the nonlinear map (16) and the cognitive dynamics (2)have the same chaotic regime as

14

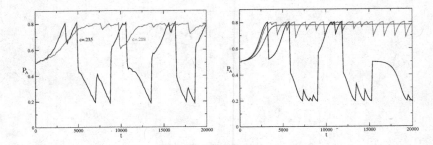

Fig. 6. Left figure: Evolution of the P_A population fraction at different value of the cooperation degree when the automata use a long decision time ($\tau_d = 250$ arbitrary time units) and a congestion threshold $P_* = .8$; the other parameters are the same of fig. 4. We observe the transition to a chaotic frustrated dynamics as the c increases, so that the automata are not able relax to a stationary cognitive state. Right figure: Dynamics of balance equations (16) for a time step $\tau_d = 250$; the transition probabilities are computed according to the Kramer's theory (9) using eq. (5) with the congestion threshold $P_A = .8$ in the utility potential (4). The different curves (blue, red, black) refer to increasing values of the cooperative degree.

when c increases. This result points out that an average approach describes the macroscopic behavior of the system as the transition to a deterministic chaotic regime, where the individual stochastic dynamics can trigger fluctuations that are then amplified up to create collective phenomena.

4. Experimental Validation Problems and Virtual Experiments

The validation of complex systems models is still a very debated problem.[1] This is not only due to the difficulty in defining relevant parameters suitable for quantitative experimental measures, but also to the intrinsic unrepeatable nature of the experiments. The usefulness of a complex system model is mainly in its capability of reproducing virtual experiments in silico that show analogous emergent properties and macroscopic states as the real systems. The virtual experiment results should be robust with respect of an entire class of microscopic interactions so that the validation process can be simplified to verify that the microscopic interactions are compatible with the experimental observations. We have performed virtual experiments with the automata gas model to study the interaction between the cognitive dynamics (2) and the physical dynamics, in simple environments. Referring to fig. 7, we consider two populations of automata moving in counterwise through two couples of parallel bottlenecks. The decision mechanism (3) is introduced to choose among the possible paths using the flows and the den-

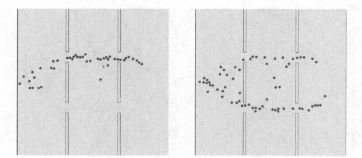

Fig. 7. Left figure: snapshot of a simulation for two automata population moving in a counterwise way through two couples of bottlenecks (the red automata move from the left to the right, whereas the green ones move from the right to the left). The automata perform a cooperative strategy with a short decision time so that a self-organized state emerges and the two populations choose different path to reach their goal. The remark there still exist few automata that do not cooperate. Right figure: a single population of cooperative automata use all the possible paths to reach the goal.

sity at the bottlenecks as relevant information. In fig. 7 (left), we show the effect of a cooperative behavior for a short decision time and a moderate population density. It is clear the formation of a self-organized state in the system when the cooperative degree overcomes a certain threshold, so that the populations choose different paths to reach their goal. Due to the probabilistic nature of the decision mechanism, there are still few automata that do not cooperate with the majority. These automata allow the system to respond efficiently to environmental changes: e.g. if one of the two population disappears, we observe a fast relaxation to a state where the automata uses all the possible paths (fig. 7 (right)). An increase of the population density strengthens the cooperative behavior and we get a completely organized state (fig. 8 (left)). But the introduction of a selfish threshold in the density destabilizes the cooperative state and a chaotic decisional regime emerges, in which the automata switch from one path to another according to the fluctuations in the density at the bottlenecks (fig. 8 (right)). The numerical simulations point out as the cognitive dynamics (3) is able to explain the formation of self-organized states in simple situations, that recall analogous phenomena experimentally observed in pedestrian dynamics.[11] At this purpose we have performed a video recording campaign during the Venetian Carnival 2007 to study the crowd dynamics in urban spaces.[15] We have developed a software to detect semi-automatically the individual dynamics from video recording by following the single head movements

 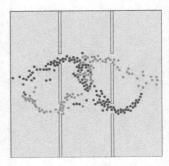

Fig. 8. Left figure: a complete coordinate state at high automata density: the system efficiency is maximal. Right figure: after the introduction of a selfish strategy the populations switch randomly from one path to another.

Fig. 9. Left figure: snapshot of the software interface developed to detect the individual dynamics from a video recording during the Venetian Carnival 2007.

taking into account the corrections due to the projection effects (fig. 9). We have analyzed the trajectories of $\simeq 400$ pedestrians moving across San Marco square from a 3 minutes video. Many people were moving forming little groups so that there is a strong correlation among their dynamics. We have divided the individuals into two main classes according to the velocity to distinguish the people that effectively cross the area from people that stop inside the area. Our analysis refers to the first class which contains approximately 70 % of the individuals ,whose trajectories are the result of the moving strategies to avoid collisions. In such a way we were able to have information on the pedestrian microdynamics in a moderate crowded situation. In fig. 10 (left) we report the distance distribution between each individual and his nearest neighbor (red curve) compare with a random distribution in the same area (blue curve). We observe as the distribution

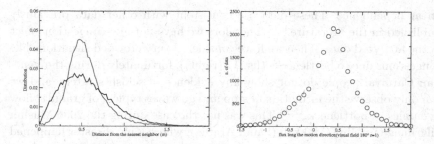

Fig. 10. Left figure: distance distribution between each individual and his nearest neighbor (Red curve) compared with a random distribution of the same number of individuals in the considered area (Blue curve). Right figure: flux distribution projected along the velocity direction for each individual in a semicircle visual space of radius $\simeq 2\ m$; the peak at .5 pedestrian per second indicates the presence of streams in the pedestrian dynamics.

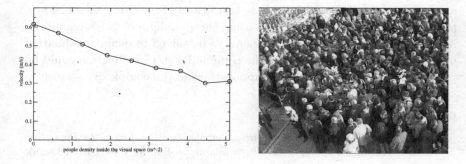

Fig. 11. Left figure: individual velocity dependence on pedestrian density in the visual space: the linear behavior means that no criticality is present in the dynamics. Right figure: snapshot of a video recording at the entrance of a bridge along Riva degli Schiavoni: stop and go density waves can be observed in the crowd.

indicate the presence of correlated structures among moving individuals and the presence of a peak at distance $\simeq .5\ m$, that can be correlated with the existence of a social space as in the automata gas model (fig. 1). The effects of a collision avoiding strategy have been studied by the flux distribution in a visual radius semicircle of dimension $\simeq 2\ m$; the results are reported in fig. 10 (right) where one can observe as the flux is peaked at a positive value denoting that people tend to move forming streams. Finally we have look for crowding effects by measuring the dependence of the individual velocity from the density in the visual space; the results are plotted in fig. 11 (left) where we do not observe any substantial deviation from a linear dependence. These data are consistent with other data previously published in the literature.[6,17] Therefore we have not any congestion effect in the analyzed data. Then we have considered very crowded situations like congestions due to bottlenecks (fig. 11 (right)), fortunately during the Venetian Carnival people do not show any evidence of selfish strategy and we can only observe the formation of stop and go waves typical of granular flow dynamics at bottleneck.[12,20] This was not the case during the 2006 Muslim pilgrimage in Mina/Makkah (Saudi Arabia), where a big disaster happened due to crowd stampede and an analysis of video recordings had pointed out the appearance of a discontinuous change in the individual behavior.[13]

5. Conclusions

In this paper we show as an automata gas, whose cognitive dynamics may realize opposite strategies, could be very useful to model the emergent dynamical states and the critical phase transitions of biological or social systems that are difficult to describe using the paradigm of Nash equilibria. Although there are still many problems to be solved to define a validation procedure, the effort of the scientific community in this direction could be very profitable for new ideas and applications of the complexity science.

References

1. Batty, M and Torrens, P, *Modeling Complexity: The Limits to Prediction* CASA Working Paper Series, **36**, available on-line at www.casa.ucl.ac.uk (2001).
2. A. Bazzani, B. Giorgini, S. Rambaldi, *Emergent Properties in an Automata Gas*, Nonlinear Dynamics Conference ENOC08, http://lib.physcon.ru, (2008).
3. Byrne D S,(1998), *Complexity Theory and the Social Sciences: An Introduction*, Routledge.
4. B. de Finetti, *Probability, Induction and Statistics*, (John Wiley & Sons New York, 1972).
5. T.A. Domencich, and V. D. Mcfadden, *Urban Travel Demand. A Behavioral Analysis*, (North Holland Publishing co., 1975).
6. J.J. Fruin, in *Engineering for Crowd Safety*, edited by R. A. Smith and J. F. Dickie, (Elsevier-Amsterdam, 1993).
7. T. van Gelder, *The dynamical hypothesis in cognitive science*, Behavioral and Brain Sciences, **21**, 615, (1998).
8. B. Giorgini, A. Bazzani, S. Rambaldi, (Eds.) *Physics and the City*, Adv. Complex Systems, **10-2**, (2007).
9. P. Hänggi, P. Talkner., M. Borkovec, *Reaction-rate theory: fifty years after Kramers* Reviews of Modern Physics, **62**, 251, (1990).

10. D. Helbing , *Traffic and Related Self-Driven Many-Particle Systems*, Reviews of Modern Physics, **73**,1067, (2001).

11. D. Helbing, L. Buzna, A. Johansson, T. Werner, *Self-organized pedestrian crowd dynamics: Experiments, simulations, and design solutions*, Transportation Science, **39**, 1, (2005).

12. D. Helbing et al, *Analytical approach to continuous and intermittent bottleneck flows*, Phys Rev Lett., **97**, 168001, (2006).

13. D. Helbing, A. Johansson, H.Z. Al-Abideen, *Dynamics of crowd disasters: an empirical study*, Phys Rev E, **75**, 046109, (2007).

14. S.P. Hoogendoorn, W. Daamen, *Pedestrian Behavior at Bottlenecks*, Transportation Science, **39(2)**,147, (2005).

15. G. Martinotti (coord.), *Individui e popolazioni in movimento*, PRIN project of the italian research ministry, (2005).

16. C.R.M. McKenzie, *Rational models as theories- not standards - of behavior* TRENDS in Cognitive Sciences **7 n.9**, 403, (2003).

17. M. Mori,H. Tsukaguchi, *A new method for evaluation of level of service in pedestrian facilities*, Transportation Research A, **21(3)**, 223, (1987).

18. L. Nadel, M. Piattelli-Palmarini, *What is cognitive science?*, Introduction to the Encyclopedia of Cognitive Science (Macmillan), (2003).

19. J. Von Neumann, *The general and logical theory of automata* in *Collected works*, **V**, Pergamon press, 288, (1963).

20. T.P.C. van Nojie, M. H. Ernst, *Kinetic theory of granula flows*, T. Pöschel and S.Lunding eds., LPN **564**, 3, (2001).

21. K. Oelschläger, *On the derivation of reaction-diffusion equations as limit dynamics of systems of moderately interacting stochastic processes*, Prob. Th. Rel. Fields, **82**, 565, (1989).

22. S. Parsons, P. Gymtrasiewicz, P., M. Wooldridge (Eds.), *Game Theory and Decision Theory in Agent-Based Systems*, Multiagent Systems, Artificial Societies, and Simulated Organizations Series , **5**, Springer, (2002).

23. K. Popper, *The propensity inetrpretation of the Porbability*, British Journal for Phylosophy of Science, (1959).

24. M.I.Rabinovich, R.Huerta, P.Varona, V.S.Afraimovich *Transient Cognitive Dynamics, Metastability, and Decision Making*, PLOS Computational Biology, **4-5**, e1000072, (2008).

25. G.J. Roederer, *On the concept of information and its role in Nature*, Entropy, **5**, 3, (2003).

26. F. Schweitzer,*Brownian Agents and Active Particles*, Springer, Berlin, (2003).

27. G. Turchetti, F. Zanlungo, B. Giorgini, *Dynamics and thermodynamics of a gas of automata*, Europhysics Letter, **78-5**, 58003, (2007).

28. J.Walhe, A.C.Bazzan, F.Klügl, M.Schreckenberg, *Decision Dynamics in a Traffic Scenario*, Adv. Complex Systems, **2**,1, (2000).

GENE-ENVIRONMENT INTERACTION: THE IMPORTANCE OF OMICS IN UNDERSTANDING THE EFFECT OF LOW-DOSE EXPOSURE

ANNAMARIA COLACCI, PAOLA SILINGARDI, MARIA GRAZIA MASCOLO, ELENA MORANDI, MONICA VACCARI

Excellence Environmental Carcinogenesis and Risk Assessment, Environmental Protection and Health Prevention Agency- Emilia Romagna, Bologna, Italy

FRANCESCA ROTONDO, STEFANIA PERDICHIZZI, SANDRO GRILLI

Alma Mater Studiorum-University of Bologna Department of Experimental Pathology-Cancer Research Section, School of Medicine, Bologna, Italy

Abstract. The risk of human exposure to environmental compounds has historically been based on the extrapolation of data from animal studies performed at high levels of exposure to single chemicals. Models used for this extrapolation include the assumption of low-dose linearity for genotoxic carcinogens and a predictable threshold for non mutagenic compounds.

This concept is increasingly questioned as the theory of hormesis is more and more supported by experimental evidence. Hormesis is a term to define the biphasic dose-response to an environmental agent that can induce stimulatory or beneficial effects at low doses and inhibitory or adverse effects at high doses. The adaptative response of living organisms to the environment poses serious questions about the reliability of the current approach for risk assessment. Some other considerations should be taken in account when defining the effects of chemicals on human health. The human population is chronically exposed to low doses of environmental pollutants. Environmental and genetic factors both play a role in the development of most diseases. Environment include any external agent, which also means food and medications. The interactions among all these aspects make the comprehension of the whole picture difficult.

The sequencing of the human genome has provided new opportunities, new technologies and new tools to study the environmental impact on living organisms and life processes. Toxicogenomics has the potential to provide all the information we need for the risk prediction in just one experimental design. Toxicogenomic technologies can be used for screening the toxic potential of environmental agents, for assessing cellular responses to different doses, for classifying chemicals on the basis of their mechanism of action, for monitoring human exposure and for predicting individual variability in the response to toxicants. We used a combination of in vitro models and in vivo studies to identify expression signatures of low-dose exposures to physical and chemical agents, highlighting hormetic effects in different target and endpoints.

1. Introduction

The risk of environmental exposure cannot be calculated only on the base of a qualitative point of view. Although we classify the physical, chemical and biological agents by the extent of damages that they can induce, the chance of a negative outcome depends on the level of exposure, i.e. duration and dose, and on the possibility that the environmental agent interacts with a target that plays a key role in the maintenance of cellular homeostasis. What is more, the response to an environmental exposure can vary according to the individual genetic susceptibility [1].

Thus, environmental and genetic factors can both play a role in the onset of many diseases and their interaction should be taken in account in the prediction of risk as well as in the prevention strategies [2, 3].

This mechanistic approach could help in understanding the effects of low-dose exposures, which can affect the general population. Low-dose exposure could have unpredictable effects, even because scientists, risk assessors and regulators do not have much literature data to rely on.

So far, the risk assessment strategy has been based on the results from animal studies, which are performed with single compounds at relatively high doses, with inbred animal strains, which are often resulted as highly susceptible to develop diseases, especially tumors [4-7]. So, the central question is: how helpful and reliable are the available data from animal bioassay in extrapolating the risk for humans? It is clear that new, more potent tools are needed to be applied in the risk assessment approach, which include highly predictive tests and a different concept of toxicology based on low-dose

2. The "exciting" hormesis theory and the low-dose toxicology

The hormesis theory is one of the most debated matters in the scientific arena. It is amazing that there is no consensus even on what hormesis means.

Hormesis is not a new term or a new concept since it had already been described by Hippocrates. Actually, hormesis is a greek-derived word meaning stimulation, and what is curious is that the root of the original word was changed to reinforce its significance. However, it was only in 1888 that some experiments in yeast gave evidence that substances vary in action depending on whether the concentration is high, medium or low. The observation became known as the Arndt-Schultz Law and had been used as the scientific base for homeopathic therapeutic approaches. That was probably the reason why hormesis is considered a theoretical phenomenon of dose-response relationships in which an agent can produce harmful biological effects at moderate to high doses and beneficial effects at low doses.

But is the cell stimulation always a beneficial effect?

Molecules that can stimulate cell proliferation play a key role in the multistep cancer process [8].

Promoting agents can trigger the cell growth through non genotoxic mechanisms, often selecting initiate cells, bearing DNA damage(s). Promoting agents include hormones, natural compounds, like phytoestrogens, drugs, man-made chemicals.

According to the most recent literature, many compounds, probably as many as 5000, can show hormetic behaviour [9]. Are all of them beneficial at low doses? Toxicologists could not agree on that.

Even if the hormesis concept acquires a slight different meaning in the fields of toxicology, biology and medicine, it could not be denied the importance of such a phenomenon in evolutionary theory. Life on earth survived despite the high concentration of mutagenic and toxic compounds and evolved because of mutations. Living organisms needed to develop mechanisms that allowed them to deal with the harsh environment that they lived in. Thus, hormesis can be considered as an adaptive response of the cell to the environmental exposure [9, 10]. However, the first (and more important) environment for a living organism is the internal system. Hormesis is part of the normal physiological function of living organisms, by which they can respond to the physiological stimuli carried out by informational molecules.

Some recent reports support the hypothesis that hormesis is a mechanisms of defence of the cell where low doses of a substance protect the cell from the toxic effects of the same substance at high doses [9].

Since it remains difficult to understand and explain hormesis, it has been proposed to differentiate the definition of hormesis from the proof of hormesis [10]. From a strictly scientific point of view, hormesis should be defined as the biphasic dose-response to an environmental agent resulting in a U, J or inverted U-shaped dose–response [8]. This implies a reviewing of the concept of the threshold dose for non genotoxic compounds and of the low-dose linearity for genotoxic carcinogens [11].

For so many years toxixologists have faced an unanswered question: what happens in the low-dose range of a dose-response curve?

As stated before, risk assessment have always been based on the results from animal bioassays performed at high-dose treatments [5]. Until a decade ago, risk was calculated on an invisible parameter, the no observed effect level

(NOEL) or no observed adverse effect level (NOAEL), meaning the level of dose where no effect was detected, that was the lowest dose tested.

In 1996 US-EPA introduced the benchmark dose (BMD) as a potential replacement of the NOAEL in risk assessment approach. The BMD is the dose that results in a predetermined level of adverse response [12]. Usually, the predetermined level is the dose that induces 10% of the expected effect. The lower confidence limit (BMDL) of the BMD is often taken as the starting point for determining allowable exposure levels. The BMD is an experimental dose, BMDL is often much lower than the lowest tested dose. This approach should allow a better assessment of any risk at exposure levels lower than experimental doses.

In any case, the BMD approach still assumes that the dose-response curve is linear in the low-dose range. If we consider the hormesis phenomenon this approach seems to be not completely appropriate.

Should the procedure for cancer risk assessment take in account the hormesis?

This would be in contrast with the paradigm for cancer risk assessment based on linearized low-dose extrapolation.

It is still difficult to draw any final conclusion on the real implication of the hormetic effect. For this reason it could be premature to include the concept of hormesis in the risk assessment, as proposed [13-15], but it would be a mistake to join the sceptical scientist in dismissing the hormesis theory as just a new faith-based religion, as the experimental evidence of hormetic effects is growing up. What is needed is a new approach to define not only the hormetic effects, but also the kind of interaction between exposure at realistic level and target molecules in living organisms.

3. The promise and dejection of soy derivatives: when low doses are not beneficial

3.1. *Phytoestrogens*

As the epidemiology studies began to accumulate evidence of the strict relationship between fruit and vegetable consumption and health benefits, the possibility to isolate the active ingredients to prevent chronic degenerative diseases became one of the main goal of biomedical research. Special attention was paid to soy derivatives since epidemiological data gave evidence for a low risk of breast and prostate cancer in Chinese and Japanese population resident in Asia whose diet is based on high intake of soy and its derivatives [16].

Soy is rich in phytoestrogens, especially isoflavones, that undergo metabolic conversion in the gut, resulting in hormone-like compounds. Isoflavones bind to the estrogen receptor, mimicking a light estrogenic activity and preventing the binding of more potent estrogens.

Genistein (4,5,7-trihydroxyisoflavone) is the main metabolite of soy [17].

The estrogenic activity of genistein has been extensively described in various tissues. Genistein binds to both the estrogen receptors (ER alfa and the ER beta subtypes), and its binding affinity is several-fold weaker than that of estradiol. When administered at low doses, genistein enhances the growth of ER positive breast cancer lines [17]. Many in vitro studies indicate a biphasic effect of genistein on human breast cancer cell lines. Due to its pleiotropic effects on several cellular processes [17] and to experimental evidences on growth inhibition of breast cancer cell lines, genistein has been proposed as a potential chemopreventive and anticancer agent for breast cancer prevention and treatment. However, genistein, as other phytoestrogens [18], exerts genotoxic effects in in vitro test systems [19] and in human cells at concentrations about 10-fold higher than that found in human after an average daily intake of soy products. Although the evidence of the range of genistein effects is far from being conclusive, it was approved as an over-the-counter soy supplement, used as an alternative to hormone replacement therapy. The toxic effects from genistein at high doses and the proliferative effects at low doses are not detected after soy intake. It is possible that soy contains other active ingredients that can counteract the estrogenic effect of genistein and protect from breast cancer. Moreover, genetic background could play an important role in response to the soy effects, since Caucasian population seem to be less sensitive [16].

For all these reasons, further investigations regarding dose-response-relationships and other appropriate aspects for extrapolation to human exposure seem necessary. Genistein, with its dualistic effect, well-known metabolic pattern and strong hormone-like properties seems, to be the best candidate to develop a panel of tests and standardize experimental approaches to investigate the hormetic effects of environmental pollutants and evaluate the risk for human exposures at low-doses.

3.2. *The alternative tests to evaluate the risk of low-dose exposure*

The need to harmonize the different carcinogens classifications in order to facilitate the global market of chemicals and to reduce the impact on human health and environment highlighted the limits of animal bioassays. The significant different carcinogenicity classification of identical chemicals indicates that assuming that tumors in animals are always indicative of human carcinogenicity may be deceptive and can induce to over-predict the carcinogenicity risk [5-7]. The likely causes of the poor predictivity of rodent carcinogenicity bioassay include: the discordance o results among rodent species, strains and gender; the often stressful routes of administration that can affect the hormonal status and the immune system; the false positive results from high dose treatments [5-7].

Cell cultures-based tests can be regarded as useful alternative approach to predict cancer risks [20]. Among the in vitro tests, BALB/c 3T3 transformation assay is a good tool to screen carcinogens, to evaluate their mechanism of action and to obtain dose-response curves [21] as well as to detect beneficial effects of promising chemopreventive agents [22]. This is the only test that is under an International pre-validation program for screening not genotoxic (promoting) agents. The information obtained in the BALB/c 3T3 model can be implemented by using cell lines representing the target organs for environmental pollutants.

3.3. *The toxicological profile of genistein in in vitro tests*

The toxicological profile and the therapeutic potential of genistein were examined by several in vitro endpoints, which are related to the multistep carcinogenesis. MCF-7 (ER-positive) and MDA-MB-231 (ER-negative) human breast cancer cell lines were chosen as eligible targets. Endpoints included cytotoxicity, cell proliferation, cell growth, anchorage independent growth, invasion. Fig 1 shows the effect of genistein on the cell clonal efficiency. At low doses (1 mM) genistein significantly enhanced the growth of MCF-7 cells At high concentrations the cloning efficiency in both ER positive and negative breast cancer cells was reduced in a dose-dependent manner. In any case, the hormetic effects at the low doses are evident. Similar results were observed in other endpoints (data not shown).

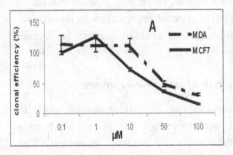 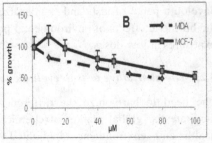

Figure 1 – Hormetic effects of genistein in breast cancer cells measured as cloning efficiency (A) and cell proliferation (B)

The GI50 value was calculated by linear regression analysis of the log(dose)/effect curve as 37.43±2.98 µM. Lower doses, ranging 1-10 µM and showing hormetic effects, induce hormone-like effects on the cell growth and can antagonize the proliferative effects of estrogens (Fig. 2).

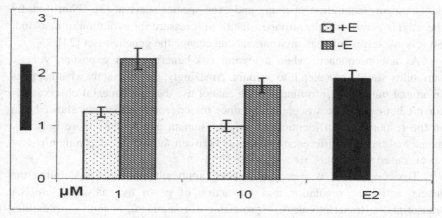

Figure 2 – Effects of genistein on breast cancer cells in the presence or absence of estrogens

4. Gene-environment interactions: towards a personalized exposure assessment

Some genetic variants generate more significant effects at low-dose of exposure. This is particularly true for genes that control the metabolic pathways of xenobiotics [1]. A genetic variant of these genes, often a polymorphism, but also deletions play a role, can originate a lower (inverse dose-effect) or higher (direct dose-effect) response with respect to the wild form of the gene [2]. This evidence has been known for a while, but it was only in the post-genomic era that many unexplainable results from experimental and epidemiological studies came to a disclosure. Surprisingly, most of these studies included kind of food or food ingredients that had failed to show the beneficial effects expected. Probably the hottest result was about beta-carotene, a carotenoid with high antioxidants properties, that contributed to increase by 18% the lung cancer in heavy smokers. More recent reports suggest that the subjects recruited for the study belonged to a population with a specific polymorphisms that increased the susceptibility to the cigarette smoking exposure and could interact with diet and chemotherapeutic agents [23]. It has also been suggested that it is the environmental exposure by itself, especially at low doses, that affects the genetic status of living organisms, rendering it less or more susceptible to the exposure. The endocrine disrupter research over the last 20 years has demonstrated that hormone-like chemicals, mimicking the physiological hormones or counteracting their function, can profoundly affect growth, development, maturation of living and even species survival [3]. Nutrition, by the way, is considered the strongest factor of pressure for evolution of mankind. So it is not surprising that environment can change the genetic asset [24].

As a consequence, when assessing risk-benefit of an exposure, genetic variability should be taken into account. Amazingly, this is exactly what happens in animal bioassay. Unfortunately, we cannot use the experimental observations for predicting genetic susceptibility, since toxicogenomics studies showed that in the response to an identical chemical, human and rodents share only few genes and that big differences exist also between the two most common rodent species used in bioassay, rat and mouse [25].

Toxicogenomics is a comprehensive approach that aims at defining the levels, activities, regulation and interaction of genes by analyzing mRNA transcripts (trancriptomics), proteins (proteomics), and metabolites (metabonomics) [26, 27]. Toxicogenomics made great strides in a different view of exposure assessment, surmounting the limits of the traditional approach looking at one chemical-one exposure in one environmental medium, giving the opportunity to investigate more realistic scenarios where human diseases are the results of multiple exposures to complex mixtures at low concentrations interacting with genetic factors. In particular, toxicogenomics-microarray gene expression analysis can give a complete picture of the effects of exposure to chemicals and other environmental stressors, providing signatures pathways of toxic injuries and even specific biomarkers of exposure, early biological response and susceptibility [26, 27]. Since the response is time and dose dependent, gene profiles can allow the comprehension of the hormesis phenomenon by highlighting specific genes involved in the response to low-doses. Toxicogenomics methods can be easily used in epidemiological and clinical studies, offering the possibility to overcome the problems related to interspecies extrapolation and to individual susceptibility. However, short-term tests, based on molecular signatures by using cell lines can be useful for guiding more expensive and time-consuming in vivo studies and for helping in the interpretation of human studies [27].

The new road map to the environmental exposure should include new technologies and methods that can improve the risk assessment. As proposed by the US Committee on Environmental Exposure Technology Development [27], a toolbox of methods, including different kind of in vivo and in vitro studies, omics approaches, functional analysis studies, which can be used alone or in combination, could provide information for different exposure in different scenarios.

By using such an approach we performed in vitro and in vivo studies and were able to profile the response t o the exposure to low doses of ionizing radiation at cellular and molecular level, by defining a set of genes whose expression is directly related to the exposure (paper in preparation).

5. Conclusion

Modern toxicology is at a crossroad. The accumulating evidence for the dualistic effects of many agents which human population is exposed to, the unpredictable outcome of some exposures that are strictly related to the genetic background of the exposed organism, the need of different tools and methodologies to evaluate the risk for human exposure to multiple agents, require a different, improved approach that takes into account insight in the mechanism underlying the risk-benefit of chemicals and other environmental agents for living organisms. This means overcoming the concept of the qualitative rating of chemical hazard by evaluating the risk at realistic low level of exposure by using omic technologies to find appropriate biomarkers of exposure and risk. This also means that the concept of the risk-benefit assessment should always be considered in the regulatory toxicology. The idea that a drug is beneficial at certain doses and has adverse side effects at higher concentration is normally accepted in medicine. A comparable approach should be advocate in toxicology. Toxicologists should have an open mind when testing and evaluating factors of possible concern, remembering the concept of Paracelsus that toxicity is just a matter of dose.

6. Acknowledgements

Authors wish to thank Prof. Adriana Izzi for the helpful and intriguing discussion about the roots of hormesis definition.

References

1. Thier R, Brüning T, Roos PH, Rihs HP, Golka K, Ko Y, Bolt HM. Markers of genetic susceptibility in human environmental hygiene and toxicology: the role of selected CYP, NAT and GST genes. Int J Hyg Environ Health. 206,149-171 (2003)
2. Garte S. Dose effects in gene environment interaction: an enzyme kinetics based approach. Med Hypotheses. 67, 488-492 (2006)
3. Edwards TM, Myers JP. Environmental exposures and gene regulation in disease etiology. Environ Health Perspect. 115, 1264-1270 (2007)

4. Gaylor, DW. Are tumor incidence rates from chronic bioassays telling us what we need to know about carcinogens? Regul Toxicol Pharmacol. 41, 128-133. (2005).

5. Grilli S. and Colacci A Is the extrapolation of carcinogenicity results from animals to humans always feasible? In: Cancer medicine at the dawn of the 21st century Eds G.Biasco and S. Tanneberger, pp 65-68 Bononia University Press (2006)

6. Knight A, Bailey J, Balcombe J. Animal carcinogenicity studies: 1. Poor human predictivity. Altern Lab Anim. 34, 19-27 (2006)

7. Knight A, Bailey J, Balcombe J. Animal carcinogenicity studies: 2. Obstacles to extrapolation of data to humans. Altern Lab Anim. 34, 29-38, (2006)

8. Fukushima S, Kinoshita A, Puatanachokchai R, Kushida M, Wanibuchi H, Morimura K. Hormesis and dose-response-mediated mechanisms in carcinogenesis: evidence for a threshold in carcinogenicity of non-genotoxic carcinogens. Carcinogenesis. 26, 1835-1845 (2005).

9. Mattson MP. Hormesis defined. Ageing Res Rev. 7, 1-7 (2008)

10. Calabrese EJ, Baldwin LA Defining hormesis. Hum Exp Toxicol, 21, 91-97 (2002)

11. Rietjens IM, Alink GM. Future of toxicology--low-dose toxicology and risk--benefit analysis. Chem Res Toxicol. 19, 977-981 (2006)

12. Travis KZ, Pate I, Welsh ZK. The role of the benchmark dose in a regulatory context. Regul Toxicol Pharmacol. 43, 280-291 (2005)

13. Jayjock MA, Lewis PG. Implications of hormesis for industrial hygiene. Hum Exp Toxicol. 21, 385-389 (2002)

14. Kitchin KT and Drane Critique of hormesis in risk assessment Hum. Exp. Toxicol. 24, 249-253 (2005)

15. Borak J, Sirianni G. Hormesis: implications for cancer risk assessment. Dose Response. 1, 443-451 (2006)

16. Peeters PH, Keinan-Boker L, van der Schouw YT, Grobbee DE. Phytoestrogens and breast cancer risk. Review of the epidemiological evidence. Breast Cancer Res Treat. 2003 Jan;77(2):171-83.

17. Bouker KB, Hilakivi-Clarke L. Genistein: does it prevent or promote breast cancer? Environ Health Perspect. 2000 Aug;108(8):701-8

18. Sirtori CR, Arnoldi A, Johnson SK Phytoestrogens: end of a tale? Ann Med. 2005;37(6):423-38.

19. Stopper, H., Schmitt, E., Kobras, K.Genotoxicity of phytoestrogens. Mutation Research 547, 139-155 (2005)

20. Knight A, Bailey J, Balcombe J. Animal carcinogenicity studies: 3. Alternatives to the bioassay. Altern Lab Anim. 34, 39-48 (2006)

21. Organisation for Economic Co-operation and Development Detailed review paper on cell transformation assays for detection of chemical carcinogens Series on testing and assessment number 31 (2007)

22. Vaccari M., Argnani A., Horn W., Silingardi P., Giungi M., Mascolo M.G., Bartoli S., Grilli S. and Colacci A. Effects of the protease inhibitor antipain on cell malignant transformation. Anticancer Res. 19, 589-596 (1999)
23. Milner JA. Molecular targets for bioactive food components. J Nutr. 134, 2492S-2498S (2004)
24. Stover PJ, Nutritional genomics Physiol Genomics 16, 161-165 (2004).
25. Sun YV, Boverhof DR, Burgoon LD, Fielden MR, Zacharewski TR Comparative analysis of dioxin response elements in human, mouse and rat genomic sequences. Nucleic Acids Res. 32, 4512-4523 (2004).
26. Waters MD, Olden K, Tennant RW. Toxicogenomic approach for assessing toxicant-related disease. Mutat Res. 544, 415-424 (2003)
27. Weis BK, Balshaw D, Barr JR, Brown D, Ellisman M, Lioy P, Omenn G, Potter JD, Smith MT, Sohn L, Suk WA, Sumner S, Swenberg J, Walt DR, Watkins S, Thompson C, Wilson SH. Personalized exposure assessment: promising approaches for human environmental health research. Environ Health Perspect. 113, 840-848 (2005).

DIFFUSION OF SHAPES

R. S. SHAW*† and N. H. PACKARD*‡§¶

*ProtoLife Inc.
6114 La Salle Ave, # 151
Oakland CA 94611

‡ European Center for Living Technology
S. Marco 2940 - 30124 Venezia, Italy

§ Santa Fe Institute
1399 Hyde Park Road
Santa Fe, New Mexico 87501

†rob@protolife.net
¶n@protolife.net

Extended objects can interact in ways not readily describable by ordinary treatments of diffusion. Here we take an informal look at some of the issues that can arise, and speculate about some possible future research directions.

1. Introduction

The movement of shapes in a crowded microscopic environment is clearly important to living systems. Any necessary motions that can be carried out by passive diffusion rather than the expenditure of ATP are a boon. In the complicated cytoplasmic environment, the diffusion coefficient will be strongly size and shape dependent, larger objects are more likely to get hung up. Over the billions of years of evolutionary history, we might expect that the geometry of interactions of shape has become highly optimized. Are there general principles describing these interactions?

These questions have received increasing attention under the rubric of "macromolecular crowding"[1,2] Chemistry in the cytoplasm, where the concentration of proteins can be upwards of forty percent, differs greatly from that in a dilute solution. Volume exclusion effects can greatly increase the effective reaction rates, as the crowding limits the possible configurations, and increases effective concentrations. Further, entropic "depletion forces" can spontaneously order shapes, simply due to the statistics of possible configurations.[3,4] Protein folding to the appropriate functional forms can be greatly enhanced in a crowded environment.[5] However, this organizational pressure comes at a cost of decreased mobility, as we can see even in a simple model.

Imagine trying trying to drive a car through the woods. As the trees become denser, it becomes harder and harder to find a path. At some density of randomly placed trees, the probability of a continuous path existing over any significant distance becomes vanishingly small. Well before this limiting value of the density, we could have difficulties, we might drive some distance only to find our way blocked, and have to back up a ways to try another route.

A continuous path through the woods must thread through the exclusion zone around each tree. If we treat the geometry of our vehicle as a simple disk (perhaps a flying saucer), this can be cast as a classical problem in percolation theory. At some density of trees, the exclusion zones will form contiguous overlapping groups across the domain, and block any path. (see figure).

How many randomly positioned holes can you drill in a steel plate before on average it falls into two pieces?[6] What density of randomly positioned cell phone towers do you need to expect to pass a message an arbitrary distance?[7] These problems are equivalent to deciding whether we can move the saucer through the woods. The question of the percolation of overlapping disks has practical applications in material science, as well as theoretical interest in scaling and universality theory.[8] Hence the density of disks of a given radius required for percolation was addressed early in the history of computer simulations[9,10] and is now known numerically to five decimal places.[11] After N disks of radius r are placed at random in an area A, the remaining free area fraction is[12] $p = e^{(-(N\pi r^2)}/A$. The critical fraction for percolation across the disks is $p_c = .32366...$

Figure 1 shows one example of 200 randomly placed "trees" in a unit square. This is near the percolation threshold for the disk radius shown at the bottom of the left hand panel. It is unclear from the point of view of the vehicle whether there exists a path through the square. For this particular member of the ensemble, there does, as we can see by drawing the exclusion zones (right hand panel). The point we would like to make with this example is that paths, when they exist, are complicated, with many forks and dead ends. A driver with a limited field of view necessarily would spend a lot of time simply *backing up* due to wrong turns. Thus

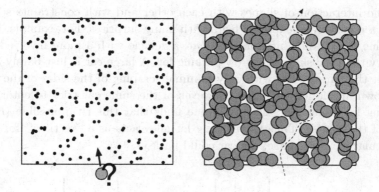

Fig. 1. (a) Can a disk go through the forest of trees? (b) Equivalent problem: can a point go through a forest of disks? Yes, if a path through for a point may be found.

even the motion of a simple disk through a random field of points naturally generates complicated paths.

Typically, in a diffusion process, the mean squared displacement of a particle increases linearly with time. This is a result of Fick's laws of diffusion, which are generally applicable when particles move randomly, as in the case of Brownian motion. Fick's theory tells us that collection of randomly moving, non-interacting point particles may be accurately described by a distribution function whose temporal behavior is governed by the diffusion equation, a simple linear field equation, with time symmetric solutions.

There has been a large amount of work on "anomalous diffusion", examining how diffusion is modified in domains containing obstacles or trapping regions, see for example Saxton.[13] Another modification of standard diffusion is to include a concentration dependent diffusion coefficient. Usually these treatments maintain the notion of a moving point particle, thus trajectories are reversible, and for example we cannot build a rectifier, although an apparent exception is motion of charged particles near appropriate charged boundaries.[14]

But if the moving objects have shape, and are densely packed, and in particular are moving in confined spaces, the simple field description starts to break down. "Shape" is a persistent set of nonlocal constraints; nonlocal in the sense that the objects are spatially extended, or "embodied". Because the objects are nonlocal, they are not readily treatable by differential operators.

The interaction of shapes with each other and with constraints seems to be a relatively unexplored field, with many surprises. For example, let's examine the possible motions of a car, a vehicle with a more involved geometry than a disk, in a narrow parking lot. A large car is just barely able to exit the parking lot, with only minimal scraping of the sides of the car, as shown at the left of Fig. 2. However, if the same car tries to enter the parking lot along the same path, one finds that this is impossible (right side of the figure). Thus the parking lot alley acts as a "rectifier" for this particular car. Of course, the car could back in. [a]

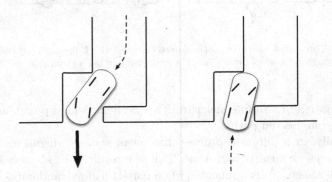

Fig. 2. Kinetic constraints and parking: It is possible to exit the parking lot, but not possible to enter the same opening. Note the difference in the differential constraints provided by the wheels in the two different cases.

For the car illustrated in Fig. 2, the turning radius - if the wheelbase of the car is R, and the steering angle is ϕ - is found by observing that two circles are generated as the vehicle moves, one by the center of the front axle, the other by the center of the rear. The first is the turning radius, $R_t = R/\sin(\phi)$. The smaller rear circle has radius $R_r = (R\cos(\phi))/\sin(\phi)$. Both can be important when maneuvering in tight spaces. When steering at $45°$, $R_r = R$, and $R_t = R\sqrt{2}$.

In the parking lot example, the tires meeting the road form a differential constraint, but we can easily arrange a similar situation occurring purely through geometry. An example is shown in figure 3, where we see that shapes in a pore may block the pore in one orientation, while passing through in another.[15]

---

[a]The car in question was a 1959 Ford station wagon, the parking lot was located in Santa Fe, New Mexico.

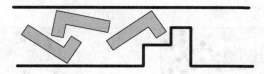

Fig. 3. Shapes blocking a pore. They need to back up and rotate to be able to go through the pore.

In both these cases, there is a branching in the configuration space, and it is possible to make a mistake resulting in an inability to move forward. There is no local clue at the fork in the road whether or not one can successfully pass through, one just has to try. The "piano mover's problem" is a classic example of the difficulties which can arise in moving a shape through constraints. It is often hard to know in advance whether a large object can be moved down a particular hallway or stairwell, and it often happens that the object becomes wedged, and one must back up a significant distance to proceed again. The general piano mover's problem is known to be computationally difficult, in fact it has been proven to be PSPACE hard.[16]

The situation is greatly complicated when there is more than one shape moving in a space. If we go driving in the woods with friends in other vehicles, a whole line of cars might need to back up if one of them gets stuck. Several puzzles are based on this fact, and they have also been proven to be computationally difficult in the same way that the piano mover's problem is difficult; one conclusion of this research is that parking lot attendants should be generously tipped for any hope of retrieving one's car.[17]

2. Membranes and rectification

To illustrate the effect geometry can have on microscopic processes, consider a model for two different sized ions flowing through a membrane: the ions are represented by hard disks, of two different sizes, and the membrane has pores illustrated in figure 4.[15,18] The pores let the smaller disks through, but the larger disks cannot go through the pore, and because of the conical asymmetry of the pore, they can enter on one side but not the other.

Figure 4(b) illustrates how a concentration differential across the membrane causes clogging, as a result of the larger disks becoming trapped inside the pore, with their escape blocked by other disks. The phenomenon is dependent on the orientation of the pores; Fig 4(a) shows how reversal of

38

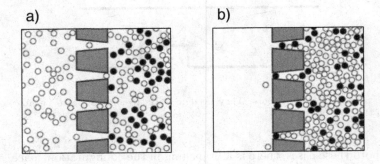

a)   b)

Fig. 4. Two sizes of hard disks diffusing through a membrane. In both cases the left side is initially empty.

the membrane allows diffusion of the small particles, given the same initial conditions as that in Fig. 4(b). The phenomenon is also crucially dependent on the size of the concentration differential. If the right side had only a low concentration, then fluctuations of the large particles could easily enable their escape from the pore. When pores become clogged, such escape becomes extremely unlikely. In the situation where a large concentration differential cause clogging of the membrane, rare fluctuations will, of course, allow particles to cross the membrane, and the concentrations on either side will equilibrate; the clogged membrane is a long-lived metastable state.

Figure 5 shows the behavior of the steady-state flux as a function of disk density for the two geometries. One side of the membrane is kept at a fixed density, the other is maintained at a vacuum. To illustrate the effect of reversing the flux through a membrane of fixed orientation, the flux and density are plotted as negative values for the first geometry. The diode action is readily apparent.

Note that this simple model naturally rectifies an alternating flux; if the concentration differential across the membrane is driven up and down fast enough to stay out of equilibrium,, the clogging effect gradually causes a buildup of concentration on the shaped side of the membrane.[18] A number of other models also display highly nonlinear relationships between driving and response, pushing harder can be counterproductive.[19-22]

The effect of lengthening the pore is shown in Fig. 6. The membrane thickness is changed while keeping the openings of the pores the same size. As the pores become longer, the membrane becomes a better and better rectifier. Flux across the membrane in the other direction is unaffected.

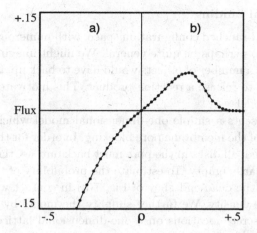

Fig. 5. Flux through the pore as a function of ρ, the disk density, for the geometries of the previous figure. Density is given as a fraction of maximum close packing, and the flux is for disks moving at unit speed on average, where the membrane is of unit width. Flux and density for panel a) are plotted as negative values.

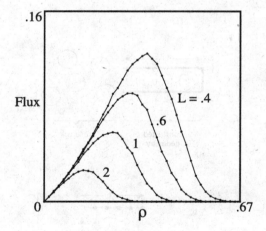

Fig. 6. Flux through the pore as a function of ρ, the disk density, for membranes of different widths. Unity is the geometry pictured in Fig 4(b). Density is a fraction of maximum close packing, and the flux is for disks moving at unit speed on average.

3. Geometrical binding

The picture of a branched configuration space, with numerous dead-ends, is simple enough to perhaps be quite general. We might imagine a number of situations where a number of objects would have to back up, to reestablish a flow, or perhaps to release a reaction product. This motivates construction of a minimal model.

Figure 7 presents a simple one-dimensional model which captures the salient features of the membrane pore blocking. In order for the block of Fig. 4(b) to be relieved, all disks in the pore must back out, i.e. the pore must be at least momentarily empty. To estimate the probability of this occurrence we can imagine the dead-end alley of Fig. 7(a) in contact with a reservoir of disks at some density. We further simplify the model by requiring that disks occupy discrete locations on a one-dimensional lattice, as shown in Fig. 7(b). The site farthest to the right represents the reservoir, and is filled at each time step with a probability ρ. If a disk comes to occupy this site, it can hop left into the interior half the time. This single reservoir site is stochastic, disks are conserved on all interior sites. The rules of motion in the interior are simply that each disk picks a direction at random, and attempts to move. If there is a free site available in that direction, it does so.

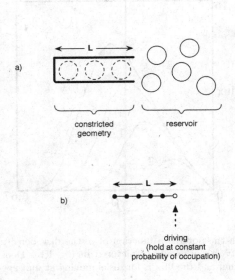

Fig. 7. Abstract representation of blocking state. A pore opens when the all sites in the lattice are unoccupied.

A master equation argument shows that the equilibrium probability of a site being occupied is the same for every site, and is equal to the feed probability ρ. The master equation does not directly tell us the probability of fluctuations. But we can see that the set of occupation patterns is "ergodic" in the sense that the all patterns occur, and with the expected probabilities. In particular, the probability of the line of Fig. 7(b) being completely empty is simply the product of the probability that each individual site is empty, or $(1 - \rho)^L$, where ρ is the feeding probability, and L is the length of the lattice. This event will clearly become exponentially unlikely as L increases.

To complete the simple model, we imagine that the "flux" F only can occur when the pore is unblocked, and that it is proportional to the concentration difference across the pore which is just ρ. So we have

$$F = \rho(1 - \rho)^L$$

The flux as a function of the feed probability ρ is plotted in Fig. 8, for a few "pores" of different length L. We see that for small imposed ρ the resulting flux is at first linear, but then declines at larger values of ρ, as the pore becomes more likely to be blocked. This effect becomes larger as the pore is lengthened. For larger values of ρ and L blockers in the pore rarely leave, and flow nearly ceases. Agreement with Fig. 6 is reasonably good.

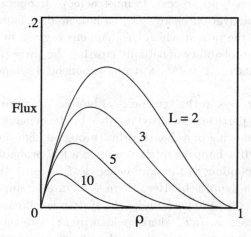

Fig. 8. "Flux" through the pore as a function of ρ, the feed probability at the entrance, for pores of different lengths.

This general picture, of a linear response near equilibrium across a membrane, and a blocking at a larger imposed concentration difference, is familiar in studies of biological membranes (see, e.g., Hille,[23] page 529). However the standard models of rectifying ion channels in membranes invoke electrostatic binding of ions in the pore to halt the flow. Our work strongly suggests that explicit energetic binding (e.g. electrostatic or covalent binding) is not necessary. The geometrical constraints imposed by the pore walls, and a dense fluid of extended objects is enough to keep blockers in place. We term this type of localization "geometrical binding".

Certainly electrostatic effects will be important to the charged ions which make up important fluxes across biological membranes. But the geometrical binding discussed above may well be significant, especially in cases where membrane pores are narrow enough so that some molecules are required to pass through in single-file fashion, as is the case with many membrane pore ion channels.[23]

4. Temperature dependence

How could we experimentally distinguish between the picture of blockers held in place by some sort of electrostatic binding, and blockers held in place by the geometrical and kinetic constraints described above? One possible way is through temperature dependence. A blocker which is localized by some sort of energetic binding must make its way over the energy barrier through a Kramers-type process. It must achieve temporarily an energy greater than an activation energy E^*, as indicated in figure 9(a). In the Kramers picture, the rate at which this happens is given by some attempt rate A, times the probability of actually crossing the barrier, which is given by an Arrhenius factor $e^{-E^*/kT}$, a strong exponential temperature dependence.

The escape process in the "geometrical binding" picture is quite different. The blocking particle is localized not by a binding energy, but by kinetic constraints. It is moving in a chaotic orbit through a high-dimensional configuration space. If it happens to chance upon a low-probability window in the space, corresponding to the pore being clear, it can leave, as schematically indicated in figure 9(b). Here there is no direct temperature dependence, although as the temperature of the system is increased, the average velocity will increase as $\sqrt{kT}$, where we identify kT with the kinetic energy $mv^2/2$. Thus we expect a relatively weak temperature dependence of this process; a particle's escape from a geometrical trap in phase space requires the system to pass through a small window, and since the time spent in the

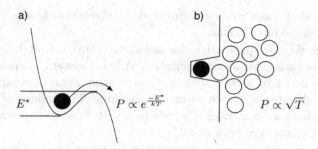

Fig. 9. (a) Energetic binding, in which the probability of the a particle escaping a local energy minimum has a strong, Arrhenius dependence on temperature. (b) Geometric binding, in which the probability of particle escape is expected to have a weaker power-law dependence (see text for discussion).

geometrical trap is proportional to velocity, we might expect the probability of escape to have a $\sqrt{T}$ dependence.

Typically though there will be other temperature dependencies in the system which will complicate the situation. For example, attractions between molecules will result in a temperature dependent overall viscosity. Also, ionic species in a real membrane pore are often in a Nernst equilibrium, which depends on the potential across the membrane and the temperature, as well as the concentrations. We can hope however that these different factors could be sorted out, and the relative importance of energetic and geometric binding ascertained in a real experiment.

5. Computation

Note that this abstract picture might well exist in chemical or catalytic systems. The "reaction coordinate" describing the passage of a chemical system from one state to another is implicitly assumed to be a single line, but there is no reason why one cannot have branched reaction coordinates. This might have significant advantages, including an automatic rectification, a tendency to drive the chemical reaction in one direction.

In a system arbitrarily close to equilibrium, there can be no forward or backward bias in a reaction rate. But with branching, we can establish a directionality with an arbitrarily small disequilibrium. Even "noise", if it is more energetic than the prevailing thermal noise, can be rectified, and used to perform useful work.[24,25] Enzyme catalysis, protein folding, and other biochemical processes necessarily involve the close interaction of shapes.

We suggest that these processes might display reaction kinetics that are effected by shape interactions.

To conclude this speculative discussion, we remark that Life in some ways is a process similar to a computation. A living system can pass though an intricate series of states, with a purpose. But living systems possess significant advantages over man-made machines. The failure of a single gate among millions can bring a laptop to a halt. From this point of view, it's amazing computers work at all.

But the "computations" performed by living systems, if we can call them such, are robust and efficient. First, there is a natural parallelism, which lends speed and noise immunity. Current computers are limited to only a few serial processors. Secondly, random numbers, in the form of fluctuations, are naturally available. Many problems in computer science require the generation of large numbers of random numbers. Often there seems no other way to perform an optimization than by a Monte Carlo or Markov chain method. The underlying problem is mathematically intractable, exponentially difficult. However randomness in living systems is not a problem. Living systems can actually use the randomness to perform large-scale parallel Monte Carlo searches.

Thirdly, constraints arise naturally in embodied systems. For instance, trying to program something like the piano mover's problem, or robotic motion planning in a crowded environment, in a conventional programming language for a serial machine is a daunting task. A numerical representation of the environment must be built, and in general one needs to check that constraints are not violated one at a time. This can become exponentially difficult. As soon as a functioning system is actually built, or "embodied", this particular difficulty disappears.

Programming of embodied systems is a key problem to be resolved in order to define and understand their computational properties and capabilities. In the examples discussed here, functional macroscopic properties (e.g. rectification) emerge as a result of the interactions of embodied microscopic elements. The functionality may be altered (or programmed, in a sense) by altering the microscopic properties of the objects and the boundaries with which they interact (e.g. membranes with pores). This sort of programming, however, is radically different from a step-by-step specification of mechanical steps of a Turing machine or any equivalent computational engine. Detailed macroscopic control is achieved without hard mechanical linkages; there is no clearly identifiable computational mechanism as Turing identified for serial computation.

There may be a parallel between motion of shaped objects in a crowded cellular environment, and the elegant forms which animals possess to glide through their surroundings. In both cases, there has evolved a natural flow of shapes, which solves the challenging problems of traversing the environment. Phrases such as "morphological computation" or "embodied computation" suggest that there may be a world of design principles out there which we now only dimly appreciate. Can we understand the principles underlying these more subtle and powerful computations? Can we build our own devices based on them?

Time will tell. Meantime, let's enjoy our drive through the woods!

References

1. R. J. Ellis, *Current opinion in structural biology* **11**, 114(February 2001).
2. A. P. Minton, *Journal of Biological Chemistry* **276**, 10577 (2001).
3. E. J. Meijer and D. Frenkel, *Phys. Rev. Lett.* **67**, 1110(Aug 1991).
4. G. Ping, G. Yang and J.-M. Yuan, *Polymer* **47**, 2564 (2006).
5. S. Schnell and T. E. Turner, *Prog Biophys Mol Biol* **85**, 235 (2004).
6. C. J. Lobb and M. G. Forrester, *Phys. Rev. B* **35**, 1899(Feb 1987).
7. L. Booth, J. Bruck, M. Franceschetti and R. Meester, *Annals of Applied Probability* **13**, 722 (2003).
8. S. F. B. I. Halperin and P. N. Sen, *Phys. Rev. Lett.* **54**, p. 2391 (1985).
9. G. E. Pike and C. H. Seager, *Phys. Rev. B* **10**, p. 1421 (1974).
10. E. T. Gawlinski and H. E. Stanley, *J. Phys. A* **14**, L291 (1981).
11. S. T. J. Quintanilla and R. Ziff, *J Phys A: Math Gen.* **33**, L399 (2000).
12. S. Chandrasekhar, *Rev. Mod. Phys.* **15**, 1 (1943).
13. M. J. Saxton, *Biophysical Journal* **66**, p. 394 (1994).
14. Z. Siwy, I. D. Kosińska, A. Fuliński and C. R. Martin, *Phys. Rev. Lett.* **94**, p. 048102 (2005).
15. N. Packard and R. Shaw, arxiv.org:cond-mat/0412626 (2004).
16. J. Reif, *Planning, Geometry, and Complexity of Robot Motion*, in *Complexity of the mover's problem and generalizations*, (Ablex Pub. Co., Norwood, N.J., 1979), Norwood, N.J., pp. 267–281.
17. G. W. Flake and E. B. Baum, *Theor. Comput. Sci.* **270**, p. 895 (2002).
18. R. S. Shaw, N. Packard, M. Schroter and H. L. Swinney, *PNAS* **104**, p. 9580 (2007).
19. R. K. P. Zia, E. L. Praestgaard and O. G. Mouritsen, *American Journal of Physics* **70**, p. 384 (2001).
20. R. L. Jack, D. Kelsey, J. P. Garrahan and D. Chandler, *Physical Review E (Statistical, Nonlinear, and Soft Matter Physics)* **78**, p. 011506 (2008).
21. M. Sellitto, *Physical Review Letters* **101**, p. 048301 (2008).
22. G. A. Cecchi and M. O. Magnasco, *Phys. Rev. Lett.* **76**, 1968 (1996).
23. B. Hille, *Ion Channels of Excitable Membranes* (Sinauer Associates, 2001).
24. R. D. Astumian and P. Hanggi, *Physics Today* **55**, p. 33 (2002).
25. C. S. Peskin, G. M. Odell and G. F. Oster, *Biophys. J* **65**, p. 316 (1993).

PART II

CLASSIFICATION AND OPTIMIZATION

CLASSIFICATION OF COLON TUMOR TISSUES USING GENETIC PROGRAMMING

FRANCESCO ARCHETTI, MAURO CASTELLI,
ILARIA GIORDANI, LEONARDO VANNESCHI

Dept. of Informatics, Systems and Communication,
University of Milano-Bicocca
20126 Milan, Italy

A Genetic Programming (GP) framework for classification is presented in this paper and applied to a publicly available biomedical microarray dataset representing a collection of expression measurements from colon biopsy experiments [3]. We report experimental results obtained using two different well known fitness criteria: the area under the receiving operating curve (ROC) and the percentage of correctly classified instances (CCI). These results, and their comparison with the ones obtained by three non-evolutionary Machine Learning methods (Support Vector Machines, Voted Perceptron and Random Forests) on the same data, seem to hint that GP is a promising technique for this kind of classification both from the viewpoint of the accuracy of the proposed solutions and of the generalization ability. These results are encouraging and should pave the way to a deeper study of GP for classification applied to biomedical microarray datasets.

1. Introduction

High-throughput microarrays have become one of the most important tools in functional genomics studies and they are commonly used to address various biological questions, like disease classification and treatment prognosis. Although cancer detection and class discovery have been often studied over the past years, no general way to work out this problem has been found yet, probably because there can be many pathways causing cancer, and a tremendous number of varieties exist. Recently, array technologies have made it straightforward to measure and monitor the expression levels of thousand of genes during cellular differentiation and response. It has been shown that specific patterns of gene expression occur during different biological states such as embryogenesis, cell development, and during normal physiological responses in tissues and cells [1]. Thus the expression of a gene provides a measure of activity of a gene under certain biochemical conditions. The key problem of evaluation of gene expression data is to find patterns in the apparently unrelated values measured. With increasing numbers of genes spotted on microarrays, visual inspection of these data has become impossible and, hence, the importance of computer analysis has substantially increased in recent years. Well-studied datasets of different phenotypes are publicly available to train and evaluate supervised pattern analysis algorithms for classification and diagnosis of unknown samples.

Therefore, there is a strong need to build molecular classifiers made of a small number of genes, especially in clinical diagnosis, where it would not be practical to have a diagnostic assay evaluate hundreds of genes in one test. In this study, we present an application of Genetic Programming (GP) [12] for molecular classification of cancer.

The paper is structured as follows: section 2.   presents a short overview of previous and related contributions. Section 3.   describes the presented GP framework and also introduces three non-evolutionary Machine Learning methods that we have used to compare with GP results. Section 4.   reports experimental results. Finally, section 5.   concludes the paper and proposes some ideas for future research.

2. State of Art

Classification is one of the most extensively studied problems in the areas of statistics, Machine Learning and databases. Many classification algorithms have been proposed, but in the last few years, researchers have started paying a growing attention to cancer classification using gene expression. Studies have shown that gene expression changes are related with different types of cancers. Many different stochastic Machine Learning methods [2] have already been applied for microarray data analysis, like k-nearest neighbors [5], hierarchical clustering [3], self-organizing maps [6], Support Vector Machines [8,13] or Bayesian networks [9]. All this different classification methods share some common issues that make classification a nontrivial task applied on gene expression data. In fact, the attribute space, or the number of genes, of the data is often huge: there are usually thousands to hundred thousands of genes present in each data set. Also, if the samples are mapped to points in the attribute space, they often can be viewed as very sparse points in a very high dimensional space. Most of existing classification algorithms were not designed with this kind of data characteristics in mind. Thus, such a situation represents a challenge for most classification algorithms. Overfitting is a major problem due to the high dimension, and the small number of observations makes generalization even harder. Furthermore, most genes are irrelevant to cancer distinction: some researchers proposed to perform a gene selection prior to cancer classification to reduce data size, thus improving the running time and remove a large number of irrelevant genes which improves the classification accuracy [2].

In the last few years Evolutionary Algorithms (EA) [24] have been used for solving problems of both selection and classification in gene expression data analysis. Genetic Algorithms (GAs) [25] have been employed for building selectors where each allele of the representation corresponds to one gene and its state denotes whether the gene is selected or not [10]. GP on the other hand has been shown to work well for recognition of structures in large data sets [14]. GP has been applied to microarray data to generate programs that reliably predict the health/malignancy states of tissue, or classify different types of tissues. An intrinsic advantage of GP is that it automatically selects a small number of feature genes during the evolution [11]. The evolution of classifiers from the initial population seamlessly integrates the process of gene selection and classifier construction. In fact, in [15] GP is used to cancer expression profiling data to select potentially informative feature genes, build molecular classifiers by mathematical integration of these genes and classify tumor samples. Furthermore, GP has been shown a promising approach for discovering comprehensible rule-based classifiers from medical data [16] as well as gene expression profiling data [17].

However, the potential of GP in cancer classification has not been fully explored. In particular, in this work, we test the power of GP using two different fitness functions: the receiving operator (*ROC*) area under curve (*AUC*) and the measure of correctly classified instances (*CCI*).

3. Material and Methods

3.1. *Dataset*

We test our application on one publicly available dataset of cancer tissues that is often used as a benchmark. *Colon cancer dataset* is a collection of expression measurements from colon biopsy samples reported in [3]. The dataset consists of 62 samples of colon epithelial cells collected from colon-cancer patients. In particular the "tumor" biopsies were extracted from tumors, and the "normal" biopsies were collected from healthy parts of the colons of the same patients. The final assignments of the status of biopsy samples were made by pathological examination. Gene expression levels in these 62 samples were measured using high density oligonucleotide arrays. Of the about 6000 genes represented in these arrays, 2000 genes were selected based on the confidence in the measured expression levels. The dataset, 62 samples over 2000 genes is available at http://microarray.princeton.edu/oncology/affydata/index.html.

In our tests we represent this dataset with an $N \times M$ matrix, where $N = 62$ (the number of samples) and $M = 2000$ (the number of genes).

3.2. *Classification methods*

In this paper we compare the performance of GP with those of three different classification techniques: Support Vector Machines (SVM), Neural Networks (in particular, we will use Voted Perceptron) and Random Forests. After a discussion of our GP framework, these methods are described below in a deliberately synthetic way, since they are well known and well established techniques. References to master those methods are quoted.

3.2.1. *Genetic programming for classification*

Candidate classifiers (individuals) that are evolved by GP are Lisp-like tree expressions built using the function set $F = \{+, *, -, /\}$ and a terminal set T composed by M floating point variables (where $M=2000$ is the number of columns in the dataset). Thus, GP individuals are arithmetic expressions, which can be transformed into binary classifiers (class "normal" for healthy tissues and class "tumor" for ill ones) by using a threshold. Here, we use two fitness functions: *ROC-AUC* and *CCI*. In the first case each classifier is evaluated by a fitness function defined as the area under the receiver operating characteristic curve [20, 21]. The ROC curve is obtained by considering 10 different threshold values uniformly distributed in the interval [0,1]. For each one of these threshold values, a point is drawn having as abscissa the false positive rate and as ordinate the true positive rate obtained by the candidate classifier using that threshold. The area is calculated using the trapezoids method. The second type of fitness function is instead obtained by fixing a particular threshold value (equal to 0.5 in this work, following [11]) and calculating the *CCI*. *CCI* is defined as the correctly classify instances rate, i.e. $CCI = (TP + TN)/N$, where *TP* indicates True Positives, *TN* specifies True Negatives and N is the number of rows in the dataset. For calculating both these fitness values during the presented GP simulations, data have not been used exactly as they are in the original dataset, but a weak gaussian noise (centered in the value itself and with a standard deviation equal to the value divided by 100) has been added to them, for improving GP generalization ability, as suggested in [29].

The other parameters we have used in our GP experiments are: population size of 200 individuals; ramped half-and-half initialization; tournament selection of size 7; maximum tree depth equal to 10; subtree crossover rate $p_c = 0.95$; subtree mutation rate $p_m = 0.1$; maximum number of generations equal to 500; furthermore, we have used generational tree based GP with elitism, i.e. unchanged copy of the best individual into the next population at each generation.

3.2.2. *Other machine learning methods*

In this paragraph we briefly describe the other machine learning methods used for our tests. For more details on these algorithms and their use, the reader is referred to the respective references quoted below.

Support Vector Machines

Support Vector Machines (SVM) were originally introduced in [18]. Their aim is to device a computationally efficient way of learning separating hyperplanes in a high dimensional feature space. In this work we use the implementation of John Platt's [19] sequential minimal optimization (SMO) algorithm for training the support vector classifier. Training a SVM requires the solution of a large quadratic programming (QP) optimization problem. SMO works by breaking this large QP problem into a series of smallest ones.

Voted perceptron

The Voted Perceptron algorithm is a margin maximizing algorithm based on an iterative application of the classic perceptron algorithm [22]. In this work we use the implementation of Breiman [23]. Voted Perceptron is not the only kind of Neural Network that we have used on the presented dataset; indeed, we have also tested a Multilayer Perceptron network, but we do not report results here, since the results that we have obtained using Voted Perceptron have a considerably better quality.

Random Forests

Random Forests is an improved Classification and Regression Trees method [27]. It works by creating a large number of classification trees or regression trees. Every tree is built using a deterministic algorithm and the trees are different owing to two factors. First, at each node, a best split is chosen from a random subset of the predictors rather than all of them. Secondly, every tree is built using a bootstrap sample of the observations. The out-of-bag (OOB) data, approximately one-third of the observations, are then used to estimate the prediction accuracy. Unlike other tree algorithms, no pruning or trimming of the fully grown tree is involved. In this work we use the implementation of Breiman [26].

4. Experimental Results

In this section we report the classification results returned by our GP framework. To obtain those results, we have performed 10 different partitions of our dataset into training and test set. For each one of these partitions, 70% of the lines of the dataset (chosen randomly with uniform probability) form the training set and the remaining 30% form the test set. Then, we have performed 100 independent GP runs for each one of these different partitions (i.e. we have performed 1000 total GP runs). Table 1 reports the obtained best, average and standard deviation of the fitness values on the test set for both fitness functions considered in this work.

Table 1: The first line reports the best fitness values found over 100 independent GP runs for each one of the 10 training-test set partitions (see text). The second line reports the average fitness over all those runs and the third line reports its standard deviation. Column one reports results obtained using ROC-AUC as fitness, while column two reports ICC results.

	ROC-AUC	CCI
best	1.0	1.0
average	0.9462	0.8941
std dev	0.0245	0.0425

The best solutions found by our GP framework correctly classify all the instances in the test set ($CCI = 1.0$) and have an ideal AUC value ($ROC\text{-}AUC = 1.0$). To have an idea of the quality of these results, we have considered the particular training-test set partitioning that has allowed us to find a solution with $CCI = 1.0$ on the test set and we have tested SVM, Voted Perceptron and Random Forests using that partitioning. These results are reported in Table 2.

Table 2: Each line reports results of a different non-evolutionary Machine Learning method. Column one reports ROC-AUC results; column two reports ICC results.

	ROC-AUC	CCI
SVM	0.9333	0.8947
Voted Perceptron	0.9083	0.7894
Random Forests	0.9083	0.8421

Comparing Table 1 and Table 2, we remark that none of the non-evolutionary methods has found neither a *CCI*, nor a *ROC-AUC* equal to 1.0 on the test set. Furthermore, if we compare the results returned by the non-evolutionary methods with the average results obtained by GP, we can see that the average values obtained by GP are comparable to (and for ROC-AUC even slightly better than) the results obtained by the best non-evolutionary method (SVM for both criteria). Finally, we point out that the GP standard deviation is quite "small" for both criteria, thus confirming that the GP behaviour is "stable" (i.e. the GP results obtained over the different runs are qualitatively similar to each other).

4.1. *The best solution found by GP*

In this section, we report a solution with $CCI = 1$ found by our GP framework. Reported as an expression in infix notation, this solution is:

```
IF(D12765/D43950/H58397*T65740*(L22214+Z15115)+(R11485
   +U20141-H69869+T84082)/U26312+((L07395-
   M14630)/(H55759+R02593)+T69026+R61332-
   M26683+L24038)*(T72889-L41268/M16827+X61123*M27635-
   R37428+M28882) > 0.5)
THEN
   Class = "tumor"
ELSE
   Class = "normal"
```

The first thing that we can remark is that GP has performed an automatic feature
selection; in fact, this solution contains only 26 over the 2000 possible genes.
These genes are briefly defined and discussed in Table 3.
For a more detailed discussion of these genes, see
http://microarray.princeton.edu/oncology/affydata/index.html.

Table 3: In this table we report a brief description of each gene contained
in the best individual found by GP. A more deep description is available at
http://microarray.princeton.edu/oncology/affydata/names.html

Gene Name	Description
D12765	Human mRNA for E1A-F
H58397	Trans-1,2-dihydrobenzene-1,2-diol dehydrogenase (human);
T65740	Single-Stranded Dna Binding Protein P9 Precursor (Mus musculus)
L22214	Receptor for adenosine a1
Z15115	DNA topoisomerase 2-beta
R11485	Proteasome subunit alpha type-5, multicatalytic proteinase complex
U20141	Human inducible nitric oxide synthase mRNA, complete cds
H69869	Clathrin Light Chain A (Human)
T84082	Er Lumen Protein Retaining Receptor 1 (Human)
U26312	Human heterochromatin protein HP1Hs-gamma mRNA, partial cds
L07395	Protein Phosphatase Pp1-Gamma Catalytic Subunit (Human)
M14630	Human prothymosin alpha mRNA, complete cds
H55759	Retinol-Binding Protein I, Cellular (Mus musculus)
T69026	60s Ribosomal Protein L9 (Human)
R61332	Ubiquitin-Activating Enzyme E1 (Human)
M26683	Human interferon gamma treatment inducible mRNA
L24038	Human murine sarcoma 3611 viral (v-raf) oncogene homolog 1 (ARAF1) gene, exons 1-16
T72889	Interferon-Inducible Protein 1-8u (Human)
L41268	Homo sapiens natural killer-associated transcript 2 (NKAT2) mRNA, complete cds
M16827	Human medium-chain acyl-CoA dehydrogenase (ACADM) mRNA, complete cds
X61123	Human BTG1 mRNA
M27635	Homo sapiens MHC HLA-DRw12 allele mRNA, beta-1 chain, complete cds
R37428	Human unknown protein mRNA, partial cds
M28882	Human MUC18 glycoprotein mRNA, complete cds

5. Conclusions and Future Work

A Genetic Programming (GP) framework for classification has been presented in this paper. It has been applied to a publicly available biomedical microarray dataset representing a collection of expression measurements from colon biopsy experiments [3]. This dataset contains 62 lines representing different tissue samples on which biopsies have been performed and 2000 columns representing gene expression levels. Each tissue is classified into "normal" if it is healthy or "tumor" if it is ill. GP experiments have been executed using two different fitness functions: the area under the receiving operating curve (*ROC*) and the percentage of correctly classified instances (*CCI*). The first one of these fitness measures is calculated using a set of threshold values (10 uniformly distributed values in the range [0,1] in this work), while the second one is obtained by fixing a predefined threshold value (0.5 in this work, following [11]). Both those fitness measures have received a noteworthy attention in past literature, but (to the best of our knowledge) they have never been studied together before.

The experimental results returned by GP have been compared with the of three non-evolutionary Machine Learning methods (Support Vector Machines, Voted Perceptron and Random Forests). They show that GP is able to find better *CCI* and *ROC* results than the best non-evolutionary method (SVM for these data).

These results are promising, even though they represent just a first preliminary step of a long term work, in which we wish to employ GP for cancer classifications in a more efficacious and structured way. Many future activities are planned. First of all, we will train our GP system in a more sophisticated way, in order to improve its generalization ability. For instance, we could use a "cross-validation like" way of dynamically handling the training set. Also, we could use more than one fitness criteria on the training set, following the idea presented in [28], where multi-optimization is shown to increment GP generalization ability in many applications. For classification, it would be particularly interesting to use both *ROC* and *CCI* during training.

One of the main limitations of this work is that we did not use any application specific problem knowledge: a "semantic" analysis of the best solutions found by GP could have helped us to generate new and possibly more effective solutions. We are currently working in this direction: we are trying to develop a sort of "application based" feature selection and in parallel we are trying to give a biological interpretation to solutions found by GP, trying to infer interesting properties.

References

1. P. Russel. Fundamentals of Genetics. Addison Wesley Longman Inc., 2000.
2. Y. Lu and J. Han. Cancer classification using gene expression data. *Inf. Syst.* 28, 4, 243-268 (Jun. 2003)

3. U. Alon, N. Barkai, D. Notterman, K. Gish, S. Ybarra, D. Mack, A. J. Levine. Broad patterns of gene expression revealed by clustering analysis of tumor and normal colon tissues probed by oligonucleotide arrays, *Proc. Nat. Acad. Sci.* USA 96, 6745–6750 (1999).
4. T.R. Golub, D.K. Slonim, P. Tamayo, M. Gaasenbeek C. Huard, J.P. Mesirov, H. Coller, M. Loh, J.R. Downing, M.A. Caligiuri, C.D. Bloomfield, and E.S. Lander. Molecular classification of cancer: Class discovery and class prediction by gene expression monitoring. *Science*, pages 531–537, (1999).
5. D. Michie, D.-J. Spiegelhalter, and C.-C. Taylor. Machine learning, neural and statistical classification. Prentice Hall, (1994).
6. A.L. Hsu, S.L. Tang, S.K. Halgamuge. An unsupervised hierarchical dynamic self-organizing approach to cancer class discovery and marker gene identification in microarray data, *Bioinformatics*, 19(16), 2131-40, 2003
7. C. Orsenigo. Gene Selection and Cancer Microarray Data Classification Via Mixed-Integer Optimization, *EvoBIO 2008:* 141-152, (2008)
8. I. Guyon, J. Weston, S. Barnhill, and V. Vapnik. Gene selection for cancer classification using support vector machines. *Machine Learning*, 46, 389–422, (2002)
9. N. Friedman, M. Linial, I. Nachmann, and D. Peer. Using Bayesian Networks to Analyze Expression Data. J. Computational Biology 7:601-620, (2000)
10. J.-J. Liu, G. Cutler, W. Li, Z. Pan, S. Peng, T. Hoey, L. Chen, and X.-B. Ling. Multiclass cancer classification and biomarker discovery using GA-based algorithms. *Bioinformatics*, 21: 2691–2697, (2005)
11. M. Rosskopf, H.A.Schmidt, U. Feldkamp, W. Banzhaf. Genetic Programming based DNA Microarray Analysis for classification of tumor tissues. Technical Report 2007-03, Memorial University of Newfoundland (2007)
12. J. Koza. Genetic Programming. MIT Press, Cambridge, MA, (1992)
13. J. C. H. Hernandez, B. Duval and J.-K. Hao. A genetic embedded approach for gene selection and classification of microarray data. *Lecture Notes in Computer Science* 4447: 90-101, Springer, (2007)
14. J.-H. Moore, J.-S. Parker, and L.-W. Hahn. Symbolic discriminant analysis for mining gene expression patterns. In L. De Raedt and P. Flach, editors, *Lecture Notes in Artificial Intelligence* 2167, 372–381, Springer, Berlin, (2001)
15. J. Yu, J. Yu, A. A. Almal, S. M. Dhanasekaran, D. Ghosh, W. P.Worzel, and A. M. Chinnaiyan. Feature Selection and Molecular Classification of Cancer Using Genetic Programming, *Neoplasia*, 2007 April; 9(4): 292–303, (2007)

58

16. C.C. Bojarczuk, H.S. Lopes, A.A. Freitas. Data mining with constrained-syntax genetic programming: applications to medical data sets, *Proceedings Intelligent Data Analysis in Medicine and Pharmacology,* (2001)

17. J.H. Hong, S.B. Cho. The classification of cancer based on DNA microarray data that uses diverse ensemble genetic programming, *Artif Intell Med.* 36:43–58, (2006)

18. V. Vapnik. *Statistical Learning Theory,* Wiley, New York, NY,(1998)

19. J. Platt. "Fast Training of Support Vector Machines using Sequential Minimal Optimization". *Advances in Kernel Methods - Support Vector Learning,* B. Schoelkopf, C. Burges, and A. Smola, eds., MIT Press, (1998)

20. C.E. Metz. Basic principles of ROC analysis. *Seminars in Nuclear Medicine,* 8, 283-298, (1978)

21. M.H. Zweig, G. Campbell. Receiver-operating characteristic (ROC) plots: a fundamental evaluation tool in clinical medicine. Clinical Chemistry, 39, 561-577, (1993)

22. F. Rosenblatt. The Perceptron: A Probabilistic Model for Information Storage and Organization in the Brain, *Cornell Aeronautical Laboratory, Psychological Review,* v65, No. 6, pp. 386-408, (1958)

23. Y. Freund and R. E. Schapire. Large margin classification using the perceptron algorithm. Proc. 11th *Annu. Conf. on Comput. Learning Theory,* pp. 209-217, ACM Press, New York, NY, (1998).

24. J. H. Holland. Adaptation in Natural and Artificial Systems. *University of Michigan Press,* Ann Arbor, (1975)

25. D.E. Goldberg. Genetic Algorithms in Search, Optimization and Machine Learning, *Addison Wesley,* (1989)

26. L. Breiman. "Random Forests". *Machine Learning* 45 (1):5-32, (2001)

27. L. Breiman, J.H. Friedman, R.A. Olshen, and C.J. Stone. Classification and Regression Trees, Belmont, California, Wadsworth International Group (1984)

28. L. Vanneschi, D. Rochat, and M. Tomassini. Multi-optimization for generalization in symbolic regression using genetic programming. In G. Nicosia et al., editor, *Proceedings of the second annual Italian Workshop on Artificial Life and Evolutionary Computation* (WIVACE 2007), (2007).

29. M. Keijzer. Scaled symbolic regression. Genetic Programming and Evolvable Machines, 5(3):259–269, (2004)

FDC-BASED PARTICLE SWARM OPTIMIZATION

ANTONIA AZZINI*, STEFANO CAGNONI** and LEONARDO VANNESCHI***

*University of Milan, Crema, Italy
** University of Parma, Italy
*** University of Milano-Bicocca, Milan, Italy

A new variant of the Particle Swarm Optimization (PSO) algorithm is presented in this paper. It uses a well-known measure of problem hardness, the Fitness-Distance Correlation, to modify the position of the swarm attractors, both global and local to single particles. The goal of the algorithm is to make the fitness landscape between each particle's positions and their attractors as smooth as possible. Experimental results, obtained on 15 out of the 25 test functions belonging to the test suite used in CEC-2005 numerical optimization competition, show that this new PSO version is generally competitive, and in some cases better, than standard PSO.

Keywords: Particle Swarm Optimizatiom, Fitness-Distance Correlation, Problem Hardness

1. Introduction

Particle Swarm Optimization (PSO)[1-3] is becoming more and more popular thanks to its effectiveness and extremely easy implementation. Many variants of the original PSO formulation have appeared to date; some of them are discussed in Section 2. We present a new variant of PSO in which a well-known measure of problem difficulty, called Fitness-Distance Correlation (FDC)[4-7] is also used to drive search. The main idea at the basis of this new PSO formulation is that swarm attractors could be generated in a more sophisticated way than simply recording the global and local best positions visited by the particles so far. In particular, our variant is designed to deal with the case in which several peaks, representing as many local optima, are located in the fitness landscape region between a given particle and the swarm global best position. In that case, we argue that using this position as an attractor may be problematic: for instance, if the particle moves towards such a global best, the presence of the intermediate peaks may contrast the attraction of the global best, especially if one of the intermediate peaks becomes the particle's local attractor. In such a situation, if we calculate the FDC using the global attractor as target (i.e. computing the particles' distances to that point), we will probably find its value to be near zero, which means that the fitness landscape is deceptive and that converging to the global best is a hard task. For this reason, we have modified the standard PSO algorithm by updating the positions of the attractors, trying to increase FDC values. As a first step of a more long-term study, we have introduced a local search algorithm into PSO which, starting from the standard attractor positions (the global and the local best positions so far) updating them trying to optimize both fitness and FDC. This local search is based on the well-known Metropolis-Hastings stochastic search algorithm.[8]

In this paper, we present this new PSO variant and compare its performance to standard PSO on the well-known benchmark suite used in the numeric optimization competition held at CEC05.[9] In the recent literature several real-parameter function optimization problems have been proposed as standard benchmark functions, like Sphere, Schwefel, Rosenbrock, Rastrigin, etc. In[9] 25 benchmark functions have been selected, to provide a fair comparison between optimization algorithms. Such benchmark suite contains, respectively, 5 unimodal and 20 multimodal functions, further divided into basic, expanded and hybrid composition functions. Twenty-two of these functions are non-separable, two are completely separable, and one is separable near the global optimum. This benchmark suite has been accepted as a more or less agreed-upon standard for testing real-parameter optimization algorithms.

The paper is structured as follows: in Section 2 we present a discussion of previous and related work on PSO and FDC. In Section 3 we describe the new PSO variant presented in this paper. Section 4 contains the experimental results we have obtained and a comparison with the ones obtained by standard PSO. Finally, Section 5 concludes the paper and offers hints for future research.

2. Previous and Related Work

2.1. *Particle Swarm Optimization and its Variants*

Many contributions have recently been dedicated to PSO, since it is easily implementable, it does not rely on gradient information, and it is able to solve a wide array of optimization problems efficiently.[10] Furthermore, as reported in,[11] PSO features reduce memory requirements and fast convergence with respect to other evolutionary algorithms (EAs). PSO basic formulation works by establishing two attractors (normally the best local and global positions so far); besides that, the swarm behavior is influenced by parameters that control global exploration and local exploitation, and try to prevent the particles from prematurely converging to local minima.[12] Starting from the work presented in,[13] several papers have been recently published which analyze and improve the performance of the PSO, trying to find the best possible values for these parameters. Several recent interesting researches in the literature describe techniques aimed at improving the performances of the PSO with different settings, which focus on the optimization of parameters such as the inertia weight, and the constriction and acceleration coefficients (see for instance[11,13–15]).

Another interesting variant of PSO original formulation consists in establishing a given "structure" (or "topology") to the swarm. Among others, Kennedy and coworkers evaluate different kinds of topologies, finding that good performance is achieved using random and Von Neumann neighborhoods.[16] Nevertheless, the authors also indicate that, selecting the most efficient neighborhood structure is in general a problem-dependent task. In,[17] Oltean and coworkers evolve the structure of an asynchronous version of the PSO algorithm. They use a hybrid technique that combines PSO with a genetic algorithm (GA), in which each GA chromosome is defined as an array which encodes an update strategy for the particles of the whole swarm. Such an approach works at macro and micro levels, that correspond, respectively, to the GA algorithm used for structure evolution, and to the PSO algorithm that assesses the quality of a GA chromosome at the macro level. The authors empirically show that the evolved PSO algorithm performs similarly and sometimes even better than standard approaches for several benchmark problems. They also indicate that, in structure evolution, several features, such as particle quality, update frequency, and swarm size influence the overall performance of PSO.[18]

Many improvements based on the conjunction of EAs and PSO have been proposed, for example considering self-update mechanisms[19] or formation of 3D complex patterns in the swarm,[20] to increase convergence speed and performance in the problems under consideration. Recently, a modified genetic PSO has been defined by Jian and colleagues,[21] which takes advantage of the crossover and mutation operators, along with a differential evolution (DE) algorithm which enhances search performance, to solve constrained optimization problems.

Other work, aimed at solving global non-linear optimization problems is presented by Kou and colleagues in.[22] They have developed a constraint-handling method in which a double PSO is used, together with an induction-enhanced evolutionary strategy technique. Two populations preserve the particles of the feasible and infeasible regions, respectively. A simple diversity mechanism is added, allowing the particles with good properties in the infeasible region to be selected for the population that preserves the particles in the feasible region. The authors state that this technique could effectively improve the convergence speed with respect to plain PSO.

To the best of our knowledge, algebraic measures of problem hardness, like the FDC, have never been used to dynamically modify the PSO dynamics before. This paper represents the first effort in this direction.

2.2. *Fitness-Distance Correlation*

An approach to measuring problem difficulty for Genetic Algorithms[4] states that an indication of problem hardness is given by the relationship between fitness and distance of the genotypes from known optima. The FDC coefficient has been used as a tool for measuring problem difficulty in genetic algorithms (GAs) and genetic programming (GP) with controversial results: some counterexamples have been found for GAs,[23] but FDC has been proven a useful measure on a large number of GA (see for example[24] or[4]) and GP functions (see[25,26]). In particular, Clergue and coworkers[25] have shown FDC to be a reasonable way of quantifying problem difficulty for GP for a set of functions. FDC is defined as the correlation coefficient between the set of the fitness values of a given sample of solutions and the set of their corresponding distances to a given global optimum. As suggested in,[4] minimization problems (like all the CEC05 test functions considered here) can be empirically classified in three classes, depending on the value of the FDC coefficient: *straightforward* ($FDC \geq 0.15$), in which fitness increases with distance, *difficult* ($-0.15 < FDC < 0.15$) in which there is virtually no correlation between fitness and distance and *misleading* ($FDC \leq -0.15$) in which fitness increases as the global optimum approaches.

3. Integration of Fitness-Distance Correlation into PSO

In this section, we describe the method that integrates the FDC measure in the PSO algorithm. We call this new PSO version FDCPSO for simplicity. Let $\mathbf{X}(t) = \{X_1, X_2, ..., X_n\}$ be the set of particles belonging to the swarm at a given time step t. Let $F = \{f_1, f_2, ..., f_n\}$ be their corresponding fitness values and let $D = \{d_1, d_2, ..., d_n\}$ be their distances to one of the swarm attractors. This attractor can be the swarm global best or the local best for a given particle. As an example, let this attractor be the global best position, that we indicate as $\mathbf{X}_{gbest}$. Then, we define: $FDC_{gbest} = C_{FD}/(\sigma_F \sigma_D)$, where: $C_{FD} = \frac{1}{n} \sum_{i=1}^{n} (f_i - \overline{f})(d_i - \overline{d})$ is the covariance of F and D and σ_F, σ_D, $\overline{f}$ and $\overline{d}$ are the standard deviations and means of F and D. For simplicity, from now on we will refer to FDC_{gbest} as "the FDC *of the* global best $\mathbf{X}_{gbest}$". As explained in Section 1, we believe that a given particle can suitably be an attractor if *its* FDC value is positive, i.e. if fitness improves as particles get closer to it. In this way, we hypothesize that the particles belonging to the swarm should "see" a smooth landscape, at least for the region of the search space that lies between their current position and the global best. For this reason, in case the current global best does not meet these requirements, FDCPSO tries to "move" the attractor to a neighboring position for which FDC is higher. In the version presented here, this "movement" of the attractors is obtained by running some iterations of a Metropolis-Hastings algorithm, where the evaluation criteria that have been considered are the FDC of the current solution under consideration and its fitness, while neighborhoods are obtained by means of a simple mutation operator that operates a Gaussian perturbation of each particle's coordinate. In other words, we have used the following function to modify the position of the swarm attractors:

```
FDC-based-Metropolis-Hastings(X_gbest) ::
    X_current = X_gbest;
    for (i = 0; i < MaxIter; i + +) do
        X_neigh = mutate(X_current);
        if (FDC_neigh ≥ FDC_current) &
            (fitness (X_neigh) ≤ fitness (X_current))
        then X_current = X_neigh;
        else
            if (rand[0, 1] ≥ α(FDC_current, FDC_neigh))
            then X_current = X_neigh;
            endif
        endif
    endfor
    return (X_current);
```

where: (1) $MaxIter$ is a prefixed constant; it has been set equal to 10 in our experiments, in order to limit the algorithm's computational complexity; (2) $\mathbf{mutate}(\mathbf{X})$ is a function that changes the value of each coordinate of its argument ($\mathbf{X}$) by means of a Gaussian perturbation centered in the value itself and with a standard deviation equal to the current value divided by 100; (3) $\mathbf{rand}[0,1]$ is a random number generated from a uniform $[0,1]$ distribution; (4) α is a function defined as $\alpha(x,y) = min\{1, \frac{y}{x}\}$, as in the standard Metropolis-Hastings algorithm.[8]

The FDCPSO algorithm can be defined as:

```
FDCPSO ::
    Initialize vectors X_i and V_i;
    X_best = X;
    X_gbest = argmin_(X_i) fitness(X_i);
    while not(termination condition) do
        for each particle 1 ≤ i ≤ n do
            V(t) = w * V(t-1) + C_1 * rand[0,1] * [X_best(t-1) - X(t-1)] +
                                          C_2 * rand[0,1] * [X_gbest(t-1) - X(t-1)]
            X(t) = X(t-1) + V(t)
            if (fitness (X) < fitness (X_best))
            then X_best = X
            endif
            if (fitness (X) < fitness (X_gbest))
            then X_gbest = X
            endif
            X_best = FDC-based-Metropolis-Hastings (X_best);
            X_gbest = FDC-based-Metropolis-Hastings (X_gbest);
        endfor
    endwhile
```

In other words, the FDCPSO works like a standard PSO, with the exception that, each time the global and local attractors are calculated, they are also transformed in the attempt to maximize their FDC, possibly without causing a worsening in fitness. However, it is important to remark that in FD-CPSO, as in any Metropolis-Hastings-based or Simulated Annealing-based algorithm, fitness can eventually worsen, according to a given probability distribution specified by the α function.

4. Experimental Results

Table 1 reports the experimental results that we have obtained by executing 25 independent runs of FDCPSO and of Standard PSO over the first 15 functions of the CEC-2005 benchmark suite for problem size equal to 10. As requested in,[9] the performance measures that we have reported are: i) the success rate (termed % in the table), i.e. the percentage of runs where the optimum has been found within the pre-specified tolerance; ii) the average number of fitness evaluations executed before finding an optimal solution divided by the success rate (FEs[9] in the table); iii) the median of the fitness evaluations executed before finding an optimal solution (Median in the table).

Table 1.   FDCPSO and Standard PSO experimental results for the first 15 functions of the CEC-2005 benchmark suite for problem size 10.

F	FDC PSO			Standard PSO		
	% succ.	FEs	Median	% succ.	FEs	Median
1	100	184.2	186	100	193.68	193
2	96	398.31	380	100	396.36	398
3	0	0	0	0	0	0
4	100	88.48	84	96	88.4549	83.5
5	0	0	0	0	0	0
6	16	14184.37	2383	8	29337.5	2347
7	0	0	0	0	0	0
8	0	0	0	0	0	0
9	0	0	0	0	0	0
10	0	0	0	0	0	0
11	0	0	0	0	0	0
12	20	6045	1209	24	4278.47	1026.5
13	0	0	0	0	0	0
14	0	0	0	0	0	0
15	0	0	0	0	0	0

Both FDCPSO and standard PSO experiments have been run setting $C_1 = C_2 = 1.49618$ and $w = 0.729844$ as proposed in.[3] By comparing results in Table 1 we can see that FDCPSO's success rate is higher than PSO's for functions 4 and 6 lower for functions 2 and 12. For all other functions, the success rate of FDCPSO is the same as for PSO. Furthermore, the number of fitness evaluations performed by FDCPSO before finding optima are generally slightly lower than the ones performed by PSO. Similar considerations can be made for results obtained on the same functions for dimensions equal to 30 and 50. These results are reported in Tables 2 and 3 respectively.

Table 2. FDCPSO and Standard PSO experimental results for the first 15 functions of the CEC-2005 benchmark suite for problem size 30.

F	FDC PSO			Standard PSO		
	% succ.	FEs	Median	% succ.	FEs	Median
1	100	521.08	514	100	543.72	537
2	96	3937.41	3780.5	100	3753.6	3732
3	0	0	0	0	0	0
4	0	0	0	0	0	0
5	0	0	0	0	0	0
6	4	180525	7221	16	27123.4	4053
7	48	1994.44	957	36	2496.6	873
8	0	0	0	0	0	0
9	0	0	0	0	0	0
10	0	0	0	0	0	0
11	0	0	0	0	0	0
12	0	0	0	0	0	0
13	0	0	0	0	0	0
14	0	0	0	0	0	0
15	0	0	0	0	0	0

Table 3. FDCPSO and Standard PSO experimental results for the first 15 functions of the CEC-2005 benchmark suite for problem size 50.

F	FDC PSO			Standard PSO		
	% succ.	FEs	Median	% succ.	FEs	Median
1	100	1220.88	1214	100	1317.08	1331
2	60	20145.11	12013	44	26926.2	11641
3	0	0	0	0	0	0
4	0	0	0	0	0	0
5	0	0	0	0	0	0
6	12	75019.44	8118	8	124919	9993.5
7	68	3651.21	2352	60	3882.22	2414
8	0	0	0	0	0	0
9	0	0	0	0	0	0
10	0	0	0	0	0	0
11	0	0	0	0	0	0
12	0	0	0	0	0	0
13	0	0	0	0	0	0
14	0	0	0	0	0	0
15	0	0	0	0	0	0

5. Conclusions

In this paper, we have modified the Particle Swarm Optimization (PSO) algorithm using a measure of problem hardness called Fitness-Distance Correlation (FDC). This PSO variant modifies the positions of the swarm attractors, both global and local to single particles. The goal of the algorithm is to allow the particles belonging to the swarm to see a smoother fitness landscape, at least between their current position and the positions of its attractors. Experimental results have been conducted on 15 of the 25 test functions belonging to the well-known CEC-2005 test suite. They suggest that this new PSO version is competitive, and in some cases better, than the standard PSO.

In the future, we plan to further improve this PSO variant by applying other measures to characterize the fitness landscape explored by the swarm and by introducing some techniques to update the values of the algorithm's constants.[11,13-15] Finally, besides using the CEC-2005 benchmark suite, we plan to test our PSO variants on some real-life optimization problems.

References

1. J. Kennedy and R. Eberhart, Particle swarm optimization, in *Proc. IEEE Int. conf. on Neural Networks*, (IEEE Computer Society, 1995).
2. Y. H. Shi and R. Eberhart, A modified particle swarm optimizer, in *Proc. IEEE Int. Conference on Evolutionary Computation*, (IEEE Computer Society, 1998).
3. M. Clerc (ed.), *Particle Swarm Optimization* (ISTE, 2006).
4. T. Jones, Evolutionary algorithms, fitness landscapes and search, Ph.D. thesis, University of New Mexico, (Albuquerque, 1995).
5. T. Jones and S. Forrest, Fitness distance correlation as a measure of problem difficulty for genetic algorithms, in *Proceedings of the Sixth International Conference on Genetic Algorithms*, ed. L. J. Eshelman (Morgan Kaufmann, 1995).
6. M. Tomassini, L. Vanneschi, P. Collard and M. Clergue, *Evolutionary Computation* **13**, 213 (2005).
7. L. Vanneschi, Theory and practice for efficient genetic programming, Ph.D. thesis, Faculty of Sciences, University of Lausanne, (Switzerland, 2004).
8. N. Madras, *Lectures on Monte Carlo Methods* (American Mathematical Society, Providence, Rhode Island, 2002).
9. P. Suganthan, N. Hansen, J. Liang, K. Deb, Y. Chen, A. Auger and S. Tiwari, *Problem Definitions and Evaluation Criteria for the CEC 2005 Special Session on Real-Parameter Optimization*, Tech. Rep. Technical Report Number 2005005, Nanyang Technological University (2005).
10. A. Esmin, G. Lambert-Torres and A. Z. de Souza, *IEEE Transactions on Power Systems* **20**, 859 (2005).
11. M. S. Arumugam and M. Rao, *Journal of Applied Soft Computing* **8**, 324 (2008).
12. A. Ratnaweera, S. Halgamuge and H. Watson, *IEEE Transactions on Evolutionary Computation* **8**, 240 (2004).
13. Y. D. Valle, G. Venayagamoorthy, S. Mohagheghi, J. Hernandez and R. Harley, *IEEE Transactions on Evolutionary Computation* **12**, 171 (2008).
14. S. Cagnoni, L. Vanneschi, A. Azzini and A. Tettamanzi, A critical assessment of some variants of particle swarm optimization, in *European Workshop on*

Bio-inspired algorithms for continuous parameter optimisation, EvoNUM'08, (Springer Verlag, 2008).

15. Z. Wu and J. Zhou, A self-adaptive particle swarm optimization algorithm with individual coefficients adjustment, in *Proc. IEEE International Conference on Computational Intelligence and Security, CIS'07*, (IEEE Computer Society, 2007).

16. J. Kennedy and R. Mendes, Population structure and particle swarm performance, in *IEEE Congress on Evolutionary Computation, CEC'02*, (IEEE Computer Society, 2002).

17. L. Diosan and M. Oltean, Evolving the structure of the particle swarm optimization algorithms, in *European Conference on Evolutionary Computation in Combinatorial Optimization, EvoCOP'06*, (Springer Verlag, 2006).

18. L. Diosan and M. Oltean, *Journal of Artificial Evolution and Applications* **1**, 1 (2008).

19. D. Srinivasan and T. H. Seow, Particle swarm inspired evolutionary algorithm (ps-ea) for multi-objective optimization problem, in *IEEE Congress on Evolutionary Computation, CEC03*, (IEEE Press, 2003).

20. H. Kwong and C. Jacob, Evolutionary exploration of dynamic swarm behaviour, in *IEEE Congress on Evolutionary Computation, CEC'03*, (IEEE Press, 2003).

21. L. Z. W. C. L. Jian, Solving contrained optimization via a modified genetic particle swarm optimization, in *Workshop on Knowledge Discovery and Data Mining, WKDD'08*, (IEEE Computer Society, 2008).

22. X. You, S. Liu and W. Zheng, Double-particle swarm optimization with induction-enhanced evolutionary strategy to solve constrained optimization problems, in *IEEE International Conference on Natural Computing, ICNC'07*, (IEEE Computer Society, 2007).

23. R. Quick, V. Rayward-Smith and G. Smith, Fitness distance correlation and ridge functions, in *Fifth Conference on Parallel Problems Solving from Nature (PPSN'98)*, (Springer, Berlin, Heidelberg, New York, 1998).

24. P. Collard, A. Gaspar, M. Clergue and C. Escazut, Fitness distance correlation as statistical measure of genetic algorithms difficulty, revisited, in *European Conference on Artificial Intelligence (ECAI'98)*, (John Witley & Sons, Ltd., Brighton, 1998).

25. M. Clergue, P. Collard, M. Tomassini and L. Vanneschi, Fitness distance correlation and problem difficulty for genetic programming, in *Proceedings of the Genetic and Evolutionary Computation Conference, GECCO'02*, ed. W. B. Langdon *et al.* (Morgan Kaufmann, San Francisco, CA, New York City, USA, 2002).

26. V. Slavov and N. I. Nikolaev, Fitness landscapes and inductive genetic programming, in *Proceedings of International Conference on Artificial Neural Networks and Genetic Algorithms (ICANNGA97)*, (Springer, Berlin, Heidelberg, New York, University of East Anglia, Norwich, UK, 1997).

A CLONAL SELECTION ALGORITHM FOR THE AUTOMATIC SYNTHESIS OF LOW-PASS FILTERS

PIERO CONCA, GIUSEPPE NICOSIA, GIOVANNI STRACQUADANIO

Department of Mathematics and Computer Science
University of Catania
Viale A. Doria 6, 95125 Catania, Italy
{conca, nicosia, stracquadanio}@dmi.unict.it

In electronics, there are two major classes of circuits, analog and digital electrical circuits. While digital circuits use discrete voltage levels, analog circuits use a continuous range of voltage. The synthesis of analog circuits is known to be a complex optimization task, due to the continuous behaviour of the output and the lack of automatic design tools; actually, the design process is almost entirely demanded to the engineers. In this research work, we introduce a new clonal selection algorithm, the *elitist Immune Programming* (EIP), which uses a new class of hypermutation operators and a network-based coding. The EIP algorithm is designed for the synthesis of topology and sizing of analog electrical circuits; in particular, it has been used for the design of passive filters. To assess the effectiveness of the designed algorithm, the obtained results have been compared with the passive filter discovered by Koza and co-authors using the Genetic Programming (GP). The circuits obtained by EIP algorithm are better than the one found by GP in terms of frequency response and number of components required to build it.

1. Introduction

The immune system consists of a complex network of process interactions, which cooperates and competes to contrast the antigen attacks. Theory of clonal selection principle hypothesizes that B-cells contrast the infections by means of a series of measures. Every being has a very large population of different B-cells within its body. In case an external entity, such as a virus or a bacterium, trespasses the body barriers, B-cells start trying to match the antigen, by means of the receptors present on their cell surface. When the receptors of a B-cell totally or partially match the antigen, the B-cell starts to proliferate in a process called *clonal expansion*. Moreover, the cloned B-cells can undergo to somatic mutations, in order to increase the affinity with an antigen: it is a Darwinian process of variation and selection, called *affinity maturation*.[1] This bottom-up behaviour has received a great attention in computer science, and it is the main source of inspiration for the emerging class of *Immune Algorithms*.[2,3]

In electronics, the design of analog circuits is an iterative process accomplished by skilled engineers. There is no CAD tool that automatically designs analog circuits starting from a set of requirements.[4] The main idea is to find a general methodology that makes effective this working flow in order to automatically design new analog circuits and speeding up the time-to-market for new devices.[5,6] In order to tackle this problem, the *elitist Immune Programming* algorithm (EIP) is introduced: it extends the *Immune Programming* (IP) algorithm[7] with the introduction of *elitism* and ad-hoc hypermutation operators for handling analog circuits. The EIP algorithm is adopted for the design of analog circuits belonging to the class of *passive filters*. Designing a Passive filter is an interesting test-bed tackled firstly by the Genetic Programming (GP) algorithm.[8–11] We have conducted several experiments in order to highlight two important aspects: firstly, how the elitism impacts the exploring and exploiting ability of the immune programming algorithm; secondly, the suitability of EIP for the automatic synthesis and sizing of analog electrical circuits. The obtained experimental results confirm that EIP outperforms the standard IP approach in terms of convergence speed and quality of the designed circuits; moreover, the new immune algorithm is able to design passive filters that are clearly better than the one discovered using GP in terms of frequency response and number of components required.

In section two we give an overview on the passive filters; in section three we describe the *elitist Immune Programming* algorithm; in section four, we report our experimental results and in section five we outline conclusions and future works.

2. Passive Filters Circuits

Passive filters are a particular class of analog circuits, which are made of passive components, such as resistors, capacitors and inductors. Given a signal, a filter leaves it unchanged in a frequency range called *pass band*; instead, in a frequency range called *stop band*, it attenuates the signal below a certain level. In the pass band, a *ripple voltage* (V_r) should be achieved; in particular, V_r is the maximum acceptable amplitude of the oscillations. In the range between pass and stop bands, called *transition band*, the filter must reduce the input signal amplitude in order to reach the desired attenuation with a very smooth behaviour. Slight deviations from an ideal behaviour are considered acceptable and they are specified by the two deviation parameters d and h.

The circuit contains a *test structure* and a *circuit core*, in this way, the same operating conditions are used for every circuit put into the core structure. The test structure is made of a signal generator (VSOURCE), a series resistance (RSOURCE), a Load Resistance (RLOAD) and a Ground link. This structure supplies three links (see Fig. 1), the first link provides the power voltage to the circuit core, which is connected to the load resistor via the second link and the third provides the connection to the ground.

In our experiments, we synthesize a passive filter with a cut-off frequency of $1KHz$ and a transition band of $1KHz$. The value for d and h were settled respectively at $0.1V$ and $10^{-4}V$ and the V_r parameter was settled to $0.03V$. The set of available values for resistors and capacitors is that of the commercial series E-24. The order of magnitude of resistors values ranges from $10^8\Omega$ to $10^{-2}\Omega$, while the order of magnitude of capacitors ranges from $10^{-1}F$ to $10^{-11}F$. For inductors there is not a similar standardization, so we have chosen values ranging from $1H$ to $10^{-11}H$ with a step size of 0.1.[9]

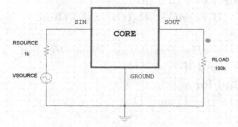

Fig. 1.   Passive Filter Circuit. It is possible to note the shunt resistance, RSOURCE, the Load Resistance, RLOAD and the power supply source, VSOURCE.

3. Immune Programming for Analog Circuit Design

IP[7] was the starting point to develop the new *elitist Immune Programming* (EIP) algorithm; the pseudo-code of the algorithm is provided in Fig.2. EIP differs from IP in several points. The following new features are introduced to effectively tackle the synthesis of topology and the sizing of analog circuits.

Firstly, the algorithm was modified with the introduction of *elitism*. At each generation g, the best solution found so far cannot be erased from the population. This strategy, already introduced in other *immune inspired* al-

```
 1: procedure EIP(D, MaxGen, Pr, Pm)
 2:     G ← 1
 3:     Population⁽⁰⁾ ← Initialize(D)
 4:     Evaluate(Population)
 5:     while G < MaxGen do
 6:         Population⁽ᴳ⁺¹⁾ ← empty
 7:         Population⁽ᴳ⁺¹⁾ ← BestCircuit[Population⁽ᴳ⁾]
 8:
        Population⁽ᴳ⁺¹⁾ ← Hypermutation[BestCircuit[Population⁽ᴳ⁾]]
 9:         i ← 0
10:         repeat
11:             if rand() < Pr then
12:                 NewCircuit ← Initialize()
13:                 Population⁽ᴳ⁺¹⁾ ← NewCircuit()
14:             else
15:                 if rand() < Pc(Circuiti) then
16:                     Population⁽ᴳ⁺¹⁾ ← Populationi⁽ᴳ⁾
17:                 end if
18:                 for j ← 1 to 2 do
19:                     if rand() < Pm(Circuiti) then
20:
        Population⁽ᴳ⁺¹⁾ ← Hypermutation[Populationi⁽ᴳ⁾]
21:                     end if
22:                 end for
23:                 i ← (i + 1) mod D
24:             end if
25:             G ← G + 1
26:         until size[Population⁽ᴳ⁺¹⁾] < D
27:     end while
28: end procedure
```

Fig. 2.   The pseudo-code of the EIP algorithm.

gorithms,[2] greatly helps the convergence of the algorithm and it overcomes the problem of IP that tends to quickly forget good solutions especially in the initial phase of the search process.   The other main difference is the application of the cloning and hypermutation operators. As in IP the chance to be cloned or mutated is driven by a parameter P_c but, in EIP, for each cloning two mutations are performed.

3.1. *Mutation operators*

The hypermutation operators operate only on the core structure; in particular, the hypermutation acts on one component, link or node at a time. All the operators take in input and return in output only feasible circuits. This design choice forces the algorithm to search in the feasible region of the solution space, and it helps the algorithm to immediately discard unfeasible solutions. Ten different mutation operators have been designed and implemented, and each of them makes a specific mutation on the circuit as described below.

ADD-SERIES. Given a circuit, it randomly selects a component and it randomly unplugs one of its terminals; successively, a new component is created and the operator connects it in series to the selected component, linking the floating terminal to the new one.

ADD-PARALLEL. It establishes a shunt connection. After a component is selected, the operator randomly creates a new component and then it links its terminals to the same nodes of the selected one.

ADD-RANDOM-COMPONENT. It randomly creates a new component that will be connected to two random nodes of the circuit.

EXPAND-NODE. This operator randomly selects a circuit node and it randomly generates a new node and a new component. Successively, it connects the new component to the previous selected node.The scope of this procedure is to easily plug in a new component into a highly linked node, or a test structure node.

DELETE-COMPONENT. This procedure tries to decrease the size of the circuit by deleting a component. It does not affect the consistency of the circuit; however, if a deletion causes damages, the operator is able to repair the circuit. An unfeasible circuit can arise due to one or more floating terminals, the unplugging of the circuit core from the test structure or the unlinking of a part of the circuit.

MUTATE-COMPONENT-VALUE. The operator randomly selects a component and it changes its value by randomly picking a new value from the set of allowed values.

COPY-COMPONENT-VALUE. The operator randomly selects a component of the circuit and it copies the value of a randomly chosen component of the same type. If there is no other similar component, it does nothing.

MUTATE-COMPONENT-KIND. This operator randomly selects a component, then it modifies the relative type and it assigns a value to the component according to the allowed set of values for the new type.

LINK-MODIFY. The operator randomly disconnects a link of a component and reconnects it to a different circuit node. Like DELETE-COMPONENT, this procedure is able to recover from unfeasible circuits.

SHRINK. The SHRINK operator scans the circuit in order to find a series or parallel connection between two or more components. It reduces the circuit size by replacing a couple of components with one equivalent component which value is as close as possible to the values of the two components. This operator greatly improves the quality of the design since it allows the automatic introduction of standard components and the reduction of the circuit size with only marginal side effects.[12]

3.2. *Fitness function*

The quality of the circuit is assessed by means of an ad-hoc objective function; it measures the distance between the curve described by a given circuit and the one described by a hypothetical ideal circuit according to the following expression:

$$F(x) = \sum_{i=100mHz}^{1KHz} [W_p(d(f_i), f_i) \times d(f_i)] + \sum_{i=2KHz}^{100MHz} [W_s(d(f_i), f_i) \times d(f_i)] \quad (1)$$

where x is a feasible circuit, f_i is the $i-th$ frequency, $d(f_i)$ is the signal deviation from an ideal behaviour and $W_p(d(f_i), f_i)$ and $W_s(d(f_i), f_i)$ are weighting factors respectively for the pass and stop band. For each frequency, the corresponding weighting factor for the pass band is determined as follows:

$$W_p(d(f_i), f_i) = \begin{cases} 0 & d(f_i) \leq V_r \\ c & V_r < d(f_i) \leq d \\ 10 \cdot c & d(f_i) > d \end{cases}$$

where V_r is the ripple voltage and d, c are experimentally obtained constants that were fixed to $d = 0.1V$ and $c = 3$. The weighting factor for the stop band term is obtained as follows:

$$W_s(d(f_i), f_i) = \begin{cases} 0 & d(f_i) \leq SBA \\ m & SBA < d(f_i) \leq h \\ 10 \cdot m & d(f_i) > h \end{cases}$$

where SBA is the desired *Stop Band Attenuation*, that was fixed to $-60dB$ and h, m are experimentally obtained constants fixed to $h = 10E-$

$5V$ and $m = 50$. It is possible to observe that the co-domain of the distance function is $[0, +\infty[$, where an ideal circuit has $F(x) = 0$. This distance function neglects small deviations from an ideal behaviour and it strongly penalizes unacceptable deviations. The fitness of each B-cell is the value of F normalized in the range $[0, 1]$ according to the following expression:

$$fitness(x_i^g) = \frac{1 - s_{f_{pf}}(x_i^g) \times m_{f_{pf}}(x_i^g)k}{\alpha} \tag{2}$$

$$s_{f_{pf}}(x_i^g) = \frac{f_{pf}(x_i^g)}{f_{pf}^{MAX}(g)} \tag{3}$$

$$m_{f_{pf}}(x_i^g) = e^{\frac{f_{pf}(x_i^g)}{k}} \tag{4}$$

where x_i^g is the $i-th$ B-cell of the population at generation g, $f_{pf}^{MAX}(g)$ is the max value of the objective function at generation g, instead k is a constant used to constraint the fitness in the range $[0, 1]$. Moreover, the fitness was scaled of $\alpha = 25\%$ in order to prevent that the worst B-cell undergoes to a complete mutation of the circuit.

4. Experimental Results

In order to assess the effectiveness of the EIP algorithm, we performed several experiments. Firstly, we compared EIP with the standard IP algorithm. We have tested these two algorithms with a population of size $D \in \{5000, 10000\}$.[13] The mutation probability parameter was settled to $P_m \in \{0.1, 0.3\}$; since P_m is the percentage of receptor mutated in the best circuit, a larger value of this parameter makes the algorithm acting as a *random search*. Finally, the replacement probability P_r and the cloning probability P_c are fixed to $P_r = 0.01, P_c = 0.2$.[7] In order to simulate the behaviour of the circuits, the tested algorithms use the NGSPICE circuit simulator. The maximum number of objective function evaluations was set to 10^7 for all the experiments and the same set of mutation operators were used in both algorithms.

It is possible to note in Tables 1 and 2 that EIP clearly outperforms the IP algorithm. For all settings, EIP shows a good convergence to near optimal solutions, instead IP produces only meaningless circuits. The scheme adopted by IP for replacement, cloning and hypermutation is not effective for this problem; at each iteration only replacement are performed and it means that IP works most likely a random search.

By inspecting the EIP results, it is possible to note that using a population of 10000 B-cells and a mutation probability $P_m = 0.1$, the algorithm

Table 1. Experimental results of the two immune algorithms. For each parameters setting, we report the *Circuit with the Lowest Fitness Function value* (CLFF) and the *Circuit with the Lowest Number of Components* (CLNC).

			CLFF		CLNC	
ALG.	d	P_m	F	Components	F	Components
IP	5×10^3	0.1	1632.04	5	1632.04	5
IP	5×10^3	0.3	1343.03	5	1343.03	5
IP	10^4	0.1	1758.54	3	1758.54	3
IP	10^4	0.3	1742.77	6	1763.77	4
EIP	5×10^3	0.1	20.5486	20	20.948	18
EIP	5×10^3	0.3	10.2221	20	11.3294	16
EIP	10^4	0.1	**0.0**	12	0.29	**10**
EIP	10^4	0.3	8.7778	18	8.78324	16

Table 2. Experimental results, a comparison of 5 independent runs of IP and EIP using the best parameter setting according to Tab.1.

RUN	ALGORITHM	f_{pf}	COMPONENTS	ALGORITHM	f_{pf}	COMPONENTS
1	IP	1542.72	3	EIP	3.93	20
2	IP	1765.63	6	EIP	16.79	20
3	IP	1658.13	6	EIP	12.62	20
4	IP	1492.22	4	EIP	0.29	10
5	IP	1497.31	3	EIP	0.0	12
	AVERAGE	1591.202	4.4	AVERAGE	**6.726**	16.4
	σ^2	118.182	1.517	σ^2	**7.591**	4.98

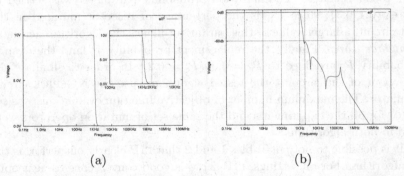

(a)   (b)

Fig. 3. The *output voltage frequency response* (a) and the *attenuation plot* (b) of the best circuit found by EIP ($F = 0.0$, 12 components).

found a circuit that perfectly matches the design requirements (Fig.3). By analyzing the circuit structure it is possible to note that is made of only 12 components that is an important aspect for the manufacturability of the fil-

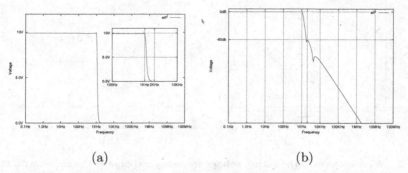

(a)   (b)

Fig. 4. The *output voltage frequency response* (a) and the *attenuation plot* (b) of the circuit with the lowest number of components found by EIP ($F = 0.29$, 10 components).

ter. Moreover, by inspecting all the circuits designed by EIP, the algorithm has found a circuit of 10 components with $F = 0.29$ (Fig.4); despite the value of the fitness function is not optimal, the circuit shows a very regular behaviour and, probably, it can be considered a good trade-off between the requirements and the manufacturability of the filter. By observing the circuits it is possible to note that they show different shapes but common building blocks; this behaviour suggests that EIP is able to find a common regular structure and, at the same time, it is able to arrange them in order to deeply explore the space of solutions. Finally, the *population-based* approach gives to the engineers not only a single solution but a set of circuits that could be inspected in order to find the one that optimally fits the design requirements.

The GP algorithm was able to find a passive filter, known as *the Campbell filter*.[8] This filter shows a very regular structure and a good symmetry, since it is built using the same building block repeated multiple times in order to form a seven rung ladder structure. The frequency response of the Campbell filter is substantially linear in pass band and the slope of the curve is very high. The two best circuits found by EIP are better then the Campbell filter for three important aspects. Firstly, in the transition band, the signal of Campbell filter shows large swings that are an undesirable behaviour instead, the EIP circuits show a very regular and smooth curve as showed in Fig.5. Secondly, the EIP circuits have only 10 and 12 components instead the Koza's circuit has 14 components, and this fact makes the EIP circuits more suitable for a real implementation. Finally, the EIP algorithm requires 10^7 fitness function evaluations to design these circuits

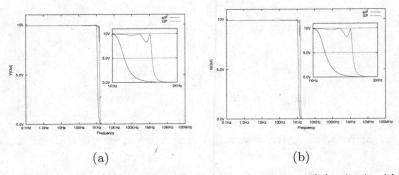

(a)   (b)

Fig. 5. A comparison of the output voltage frequency response of the circuit with the optimal fitness function value (a, $F = 0.0$, 12 components) and the one with the lowest number of components (b, $F = 0.29$, 10 components) found by EIP with the *Campbell filter*.[8] It is possible to note that in the transition band the *Campbell filter* has not a regular behaviour instead the EIP circuits have a regular and smooth curve.

instead the GP algorithm requires 1.5×10^7 evaluations; this experimental result proves that the immune algorithm, for this design problem, is more efficient than GP on the same objective function.

5. Conclusions and Future Works

In this research work, we have introduced a new *immune algorithm*, called ELITIST IP, for the synthesis of topology and sizing of analog electrical circuits. The algorithm extends the IMMUNE PROGRAMMING approach with the introduction of *elitism* and *ad-hoc operators* for handling analog circuits.

The experimental results confirms that EIP clearly outperforms the standard IMMUNE PROGRAMMING approach in terms of quality of the circuits and convergence rate. The analysis of the EIP circuits shows that the algorithm is able to synthesize analog circuits with excellent frequency responses, having small swings, high inclination and a good shape regularity. The comparison with the *Campbell filter*, a passive filter discovered using GENETIC PROGRAMMING, shows that EIP is able to find a better circuit in terms of regularity in transition band and number of components required.

Starting from these results, there are two major fields that we are investigating. Firstly, we are extending the EIP algorithm in order to use a selection strategy based on the *Pareto Optimality* criterion; using this approach, it is possible to explicitly introduce different design requirements, such as the number of components and the frequency response, and leaving to the algorithm the automatic discovering of optimal trade-off.[14] Finally, we are designing an improved EIP that is able to synthesize the topology and the sizing of active filters;[15] this last task is a visionary research topic since there is not an automatic approach for the design of these analog circuits and it could be an important step to dramatically decrease the time-to-market required for analog circuits.

References

1. A. Abbas, A. Lichtman and J. e. a. Pober, *Cellular and molecular immunology* (W. B. Saunders, 2000).
2. V. Cutello, G. Nicosia and M. Pavone, *Proceedings of the 2006 ACM symposium on Applied computing* , 950 (2006).
3. V. Cutello, G. Nicosia, M. Pavone and J. Timmis, *Evolutionary Computation, IEEE Transactions on* **11**, 101 (2007).
4. M. Streeter, M. Keane and J. Koza, *Proceedings of the Genetic and Evolutionary Computation Conference* , 877 (2002).
5. J. Koza, F. Bennett III, D. Andre, M. Keane and F. Dunlap, *Evolutionary Computation, IEEE Transactions on* **1**, 109 (1997).
6. N. Kashtan and U. Alon, *Proceedings of the National Academy of Sciences* **102**, 13773 (2005).
7. P. Musilek, A. Lau, M. Reformat and L. Wyard-Scott, *Information Sciences* **176**, 972 (2006).
8. J. Koza, F. Bennett III, D. Andre and M. Keane, *Computer Methods in Applied Mechanics and Engineering* **186**, 459 (2000).
9. J. Koza, L. Jones, M. Keane and M. Streeter, *Genetic Programming Theory and Practice II* (2004).
10. J. Grimbleby, *Circuits, Devices and Systems, IEE Proceedings [see also IEE Proceedings G-Circuits, Devices and Systems]* **147**, 319 (2000).
11. G. Alpaydin, S. Balkir and G. Dundar, *Evolutionary Computation, IEEE Transactions on* **7**, 240 (2003).
12. T. Dastidar, P. Chakrabarti and P. Ray, *Evolutionary Computation, IEEE Transactions on* **9**, 211 (2005).
13. J. Koza, *Genetic Programming III: Darwinian Invention and Problem Solving* (Morgan Kaufmann, 1999).
14. A. Subramanian and A. Sayed, *Automatic Control, IEEE Transactions on* **49**, 149 (2004).
15. M. El-Habrouk, M. Darwish and P. Mehta, *Electric Power Applications, IEE Proceedings* **147**, 403 (2000).

AN EVOLUTIONARY PREDICTIVE APPROACH TO DESIGN HIGH DIMENSIONAL EXPERIMENTS

DAVIDE DE MARCH

Department of Statistics, University of Florence
Viale Morgagni 59, 50134 Florence, Italy

European Centre for Living Technology
San. Marco 2940, 30124 Venice, Italy
davidedemarch@unive.it

MICHELE FORLIN

Department of information and telecommunication technology, Uuniversity of Trento
Via Sommarive 14, 38000 Povo (TN), Italy

DEBORA SLANZI

Department of statistics, University Ca' Foscari of Venice
Cannaregio 873, 30121 Venice Italy

European Centre for Living Technology
San Marco 2940, 30124 Venice, Italy

IRENE POLI

Department of statistics, University Ca' Foscari of Venice
Cannaregio 873, 30121 Venice Italy

European Centre for Living Technology
San Marco 2940, 30124 Venice, Italy

Abstract

The high dimensional nature of the experimental research space is increasingly present in a variety of scientific areas, and characterizes at the moment most of biochemical experimentation. High dimensionality may pose serious problems in design the experiments because of the cost and feasibility of a large number of experiments, as requested by classical combinatorial design of experiments. In this work we propose a predictive approach based on neural network stochastic models where the design is derived in a sequential evolutionary way. From a set of neural networks trained on an initial random population of experiments, the best predictive net is chosen on a different validation set of experiments and adopted to predict the unknown space. The experiments which satisfy a predefined optimality criterion are then chosen and added to the initial population to form a second generation of experiments. The algorithm then proceeds through generations with a stopping rule on the convergence of the result. This approach, that allows to investigate large experimental spaces with a very small number of experiments, has been evaluated with a simulation study and shows a very good performance also in comparison with the genetic algorithm's approach.

Key words: Design of experiments; combinatorial complexity; high dimensionality; predictive neural networks designs.

1. Introduction

Several areas of experimental research are increasingly characterized by large sets of parameters that can affect the result of the experimentation. The increasing availability of biological and physical information components, the more powerful technology for conducting the experiments and the very challenging experimental questions pose the problem of searching in very high dimensional spaces. In biochemical experimentation the huge libraries of candidate chemical compounds now available, the different ways one can create compositions, and the different laboratory protocols produce a dramatic expansion of the number of parameters to control.

Designing these experiments and formulating an optimization strategy is becoming a very difficult and challenging problem. If we adopt the classical combinatorial design of experiments (Montgomery, 2005; Cox & Reid, 2000) we should be asked in fact to conduct a huge set of experiments: with just 10 molecules at five concentration levels, to be mixed in an experiment, we should conduct 5^{10} experiments, namely 9765625 different experiments. When the experiments are expansive and time consuming, as in the biochemical or pharmaceutical studies, it is essential to design a strategy of experimentation that, involving just a small number of experiments can be informative on the questions posed to the experimentation.

In this paper we introduce an evolutionary model-based design of experiments, the Predictive Neural Network–design (PNN-design).

Avoiding dimensionality reduction procedures that may mislead the search and hide fundamental components or interactions, we develop a non-standard approach based on evolutionary statistical procedures where the design is not selected *a priori*, but evolved in a sequential way and according to the information achieved from the models. More specifically, we build a stochastic neural network model on a very small set of experiments, and adopt this model to predict the results of all the experiments in the unknown space. The best experiments, according to a predefined optimality criterion, are then chosen to increase the initial set of experiments on which we build a different neural network model. This model learns from the new set of data and develops more accurate predictions on the rest of the space. The procedure continues till a convergence optimality value is reached.

The approach is adaptive, since the model changes with the increasing set of experiments, and evolutionary, since the search strategy evolves with the information achieved at each step of the procedure. We evaluate the performance of the approach with a simulation study where we choose a very simple structure of the network and select a very small set of experiments. In this simulation the PNN-design shows an excellent performance in reaching the optimality region also in comparison with the genetic algorithm design (GA-design) proposed as a benchmark in the analysis.

This novel methodological approach to design experiments has been mainly motivated for biochemical experimentation dedicated to the study of the emergence of synthetic cells (Theis, et al.; Forlin, et al.), and will be further applied to this problem.

In the paper Section 2 describes the predictive neural network algorithm to design the experiments, namely the PNN-design; Section 3 presents a simulation analysis, conducted to evaluate the performance of the algorithm and to develop a comparison with the genetic algorithm design here considered as the best competitor for designing high dimensional space. Concluding remarks are presented in Section 4.

2. The predictive approach to design the experiments

To address the problem of selecting a very small set of experiments for uncovering the experimental composition that gives rise to the best solution in a large and complex space we propose a predictive algorithmic approach based on a stochastic multilayer feedforward network.

Initially we consider a very small population of experiments, selected in a random way from a huge space of candidates that represent the experimental space of the problem. Denoting $(e_1, e_2, \dots, e_n)$ the set of all the experiments we select randomly the subset $(e_1, e_2, \dots, e_m)$ where m<<n. Each experiment is

identified by a point in a d-dimensional space where d is a high value. The structure of each experiment may be described by a vector of elements, $e_i = (\acute{x}_1, x_2, \ldots, x_p)$, i=1,...,m, identifying the parameters that can affect the result of the experiments. These set of parameters include composition variables (compounds in different concentration levels) and process variables (temperature, reaction times, pH levels, ...). The quality of each experiment is measured by a defined response function that is to be maximized in order to optimize the system.

This first random population of experiments is thus conducted and the response of the process is measured. In order to explore the search space with a predictive selective tool, we then build a neural net model for this initial population data set. We choose a very simple neural network with a three layers topology and a feed-forward dynamics. The connections of the network are estimated on a subset of this population, identified as the training set of data, and predictions are derived on the remaining set of experiments, identified as the validation set. All the evaluations are performed with the free software R (http://cran.r-project.org/), adopting specifically the package *nnet*, with sigmoid activation functions that connect the input layer and the hidden layer and linear activation functions connecting the hidden layer and the output layer.

The accuracy of the predictions selects the best neural network model to explore the unknown space of candidate experiments and select a population of experiments that (virtually) satisfy a defined optimality criterion. This new population of experiments is then added to the initial random population to form the second generation of experiments that are to be conducted in order to achieve their responses. We can now study a data set that includes the first and the second populations of data and build on this larger set the second predictive neural network model to explore the experimental space and uncover the subsequent population that satisfies a predefined optimality criterion. The algorithm then continues for different generations with a stopping rule identified on a convergence value reached for a set of generations.

This Predictive Neural Networks Design, PNN-Design, selects the experimental design in an evolutionary and adaptive way: the design, involving a small number of experiments, is evolved through different generations via a neural network model that, learning from data and predicting the unexplored space, can achieve information to improve the design and intelligently adjust the way to the optimality region of the experimental space.

3. A simulation study

To evaluate the performance of this predictive approach and develop comparisons with other search strategies we build a simulation platform. A stochastic generating process, described by a polynomial regression model with high order interactions between variables and a white noise error term is assumed to initiate the study. This model produces a landscape over the experimental space with a gaussian shaped distribution and a global maximum corresponding to a single experiment.

A random population of 30 experiments is then selected and their response is achieved by the generating process. On these resulting data we build a three layers neural network, selecting 80% of the experiments for training the network and 20% to evaluate the accuracy of the prediction. We then simulate 100 networks assuming a three layer structure with 2 and 3 nodes in the hidden layer. From a set of real experiments conducted to study the vesicles formation in amphiphilic systems, (as described in Theis et al. (2008) and Forlin et al. (2008), we assume a 16-dimensional experimental space, building a 16 variables input layer, with 6 level for each dimension, and 1 variable output layer. We then estimate the connections of all these networks and derive as a measure of prediction accuracy the Predictive Error

$$PE = \sqrt{\frac{\sum\limits_{i=1}^{s} \left(Y_i - \hat{Y}_i\right)^2}{s}}$$

where Y_i denote the results of experiments achieved with the simulation, $\hat{Y}_i$ denote the predictions obtained by the network estimated on the training set and s represents the size of the validation set. The network that presents the minimum predictive root mean square error is then considered as the best predictive network and it is adopted to predict the response of all the experiments of the search space. From these evaluations we select the best 30 experiments (with the highest results according to the optimality criterion proposed), and join these experiments to the first generation. The algorithm then proceeds considering this new generation of experiments, that include the initial random set and the best predicted set achieved with the network model. This new set of experiments, that is supposed to be more informative with respect to the problem studied, provides data for a new set of networks to predict the experimental space. The procedure than continues with more and more informative experiments in the way to the optimality region. In this simulation study we continue to the 10th generation, where a convergence response seems to be reached. We present in the following figure the results, and the evolution of the results through generations, that we achieved in this study. To evaluate the performance of the PNN-design we also develop the evolutionary design based on a genetic algorithm, the GA-design, that is regarded a very good competitor for high dimensional problems. This algorithm is a simple genetic algorithm, adjusted to the mixture structure of the experiments. The details of this algorithm are described in Forlin et al. (2008).

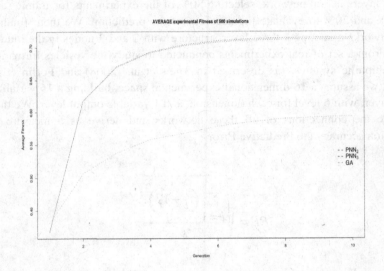

Fig. 1 The average response of the system through generations of experiments achieved with the Predictive Neural Network Design (assuming 2 and 3 hidden nodes topology), confronted with the Genetic Algorithm Design.

Observing the Fig. 1 we can follow the evolution of the average response of the system in the search of the optimality region of the experimental space: from the first generation, that is random for both the designs here considered, we can observe an excellent behaviour of the PNN-design, that can reach very high values of the average response in very few generations of the algorithm. This behaviour is achieved with a simple topology of the neural network, that involves just three feed forward layers, and a limited structure of the layers, that includes 2 and 3 hidden nodes. The excellent performance of the PNN-design is shown also in comparison with the GA-design: the difference is evident from the second generation and then increases maintaining a large gap between the approaches. These results, that have been achieved under the assumption of a Gaussian probabilistic structure of the data generating process, seem to be confirmed also under asymmetric distribution assumptions. Initial simulation studies on different probability distributions confirm these results and also inform on the robustness of the approach.

4. Conclusion

In this paper we introduced an algorithmic approach to design experiments in high dimensional settings, the PNN-design. This approach is based on a class of predictive neural networks derived on small set of data and able to explore the experimental space and uncover the experiment, or the set of experiments, that are regarded optimal with respect to the problem studied. In a simulation analysis the PNN-design shows a very impressive performance, reaching in few generations of the algorithm the region of optimality. The PNN-design also reached its objective in a faster and more effective way than the Genetic Algorithm design here considered as a benchmark. Further simulation studies, that will be soon completed, have also shown the robustness of the approach with respect to the data generation process adopted for simulation.

This novel evolutionary approach will be applied to real experiments dedicated to the formation of synthetic components of an artificial cell. These experiments will be conducted at the LivingTech Laboratory within the DICE project.

Acknowledgements

This work was supported by the EU integrated project Programmable Artificial Cell Evolution (PACE) in FP6-IST-FET, Complex Systems Initiative and by the Fondazione di Venezia (DICE project), grateful acknowledged. The authors would like to thank also the European Centre for Living Technology (www.ecltech.org) for providing opportunities of presentation and fruitful discussion of the research. Thanks to the colleagues of the University of Venice and of ProtoLife which gave valuable suggestions to this work.

References

1. M. A. Bedau, A. Buchanan, G. Gazzola, M. M., Hanczyc, T. Maeke, J. S. McCaskill, I. Poli, N. Packard, Evolutionary Design of a DDPD Model of Ligation. Artificial Evolution, 3871: 201-212, 2005.
2. J. Cawse, Experimental design for combinatorial and high throughput materials development. John Wiley and Sons, Hoboken, NJ, 2003.
3. D. M. Cox, N. Reid, The theory of the design of experiments. Chapman & Hall, London, 2000.
4. M. Forlin, I. Poli, D. De March, N. Packard, G. Gazzola, R. Serra, Evolutionary experiments for self-assembling amphiphilic systems, Chemometrics and Intelligent Laboratory Systems, 90 (2), 153-160, 2008.
5. Hornik K., Stinchcombe M., e White H. Multilayer feedforward networks are universal approximators. Neural Networks, 2 (5), 359-366, 1989.
6. T. Minerva, S. Paterlini, I. Poli, A genetic algorithm for neural network design (GANND), Economics & Complexity, 2000.
7. D.C. Montgomery, Design and Analysis of Experiments, 2005, Wiley & Son, New York, NY.
8. I. Poli, R. D. Jones, A neural net model for prediction, Journal of the American Statistical Association, vol. 89, 1994.
9. M. Theis, G. Gazzola, M. Forlin, I. Poli, M.M. Hanczyc, N. Packard, M. Bedau: Optimal Formulation of Complex Chemical Systems with a Genetic Algorithm, COMPLEXUS, 2008.
10. J. Zupan, J. Gasteiger, Neural networks in chemistry and drug design, Wiley-VCH, Weinheim, 1999.

PARTICLE SWARM FOR PATTERN MATCHING IN IMAGE ANALYSIS

L. MUSSI*

Dipartimento di Matematica e Informatica, Università degli Studi di Perugia,
Via Vanvitelli 1, I-06123 Perugia, Italy
* E-mail: mussi@dipmat.unipg.it

S. CAGNONI

Dipartimento di Ingegneria dell'Informazione, Università degli Studi di Parma,
Viale G. Usberti 181a, I-43100 Parma, Italy
E-mail: cagnoni@ce.unipr.it

This work presents a preliminary investigation on the use of a Particle Swarm Optimization (PSO) algorithm variant for Pattern Matching in image analysis. Providing each particle with its own target and having them organized with the classical Von Neumann topology is shown to be a feasible way to obtain a swarm able to locate a pattern on a digital image.

Some preliminary tests on synthetic images show the effectiveness of the modified swarm algorithm, highlighting its insensitivity to basic transforms like mirroring, scaling and perspective deformations of the pattern.

Keywords: Swarm Intelligence; PSO; Pattern Matching; Image Analysis.

1. Introduction

Particle Swarm Optimization (PSO) is a simple but powerful optimization algorithm, introduced by Kennedy and Eberhart in 1995.[1] PSO searches for the optima of a function, termed *fitness function*, following rules inspired by the behavior of flocks of birds looking for food. Many variants of the basic algorithm have been developed,[2] some of which have focused on the different neighborhood topologies according to which the swarm can be organized.[3]

In the last decade PSO has been successfully applied to many problems in several fields,[4] image analysis being one of the most frequent ones. In fact, image analysis problems can be often reformulated as optimization tasks, in which an objective function, directly derived from the physical features of the problem, is optimized. A different approach considers PSO, or more in general swarm intelligence paradigms, not only as a way to "tune" the parameters of another algorithm, but directly as part of the solution: in this case one can assume the swarm to fly over the image to detect points or regions of interest.[5-9]

In this paper we propose some tweaks which can be applied to standard PSO to let the swarm perform pattern/template matching. We call this modified PSO algorithm "Pattern Matching - PSO" (PM-PSO).

2. PM-PSO

The basic idea on which we rely to create a pattern-matching swarm assumes that individuals are free to move around the image looking for clues of the presence of the pattern. In the most basic case, a single clue could be a pixel which has a color very similar to one of the template's. However, isolated clues are not robust indicators, so the swarm must eventually converge on a clean-cut area within which most pixels have colors similar to the pattern's. This is still not enough. The relative position of all pixels inside this area must resemble the original spatial organization of the template. In other words, the swarm must consider both pattern colors and their topology to detect the presence of the template to be matched.

Instead of encoding this in a common fitness function, each particle could seek for a different small element of the template while the neighborhood topology of the swarm could be used to verify whether the relative positioning of the pixels under consideration correspond to the template's.

Starting from the basic PSO algorithm with the individuals organized according to a Von Neumann topology, we have provided each particle with the capacity to "learn" its own specific target. Moreover, we have modified the velocity update function to let the swarm spread and self-organize as regularly as possible over the whole area characterized by high fitness values.

The following subsections provide details about the swarm training and the modifications we have applied to the fitness function and the velocity update equation.

2.1. *Swarm "Training"*

To set the "personal" target of each individual in the swarm, the whole lattice is initially superimposed on the template to be matched. All particles must be uniformly distributed to cover the whole template and must be equally spaced along each direction, such that each particle represents its neighborhood barycenter. Under these conditions, each individual samples the template at its current position and memorizes the color pT (personal target) to hunt. After this initialization step, the swarm is usually randomly spread on the image to be analyzed and all particles are allowed to move until certain stopping criteria are met.

2.2. Fitness Function

The fitness function associated with each particle, which we call "total fitness" tF, is composed by many terms. This derives from the idea that the fitness of an individual must not just represent an indicator of its current state, but must also comprise a social term that makes it "fitter" when also its neighbors are "fit", as well as some other indices which measure the degree of order which characterizes its neighborhood.

The first term permits an individual to seek for its personal target; we call it "personal fitness", defined as:

$$pF(t) = e^{-3 \cdot d_{YUV}(I(\boldsymbol{X}(t)), \boldsymbol{pT})} \tag{1}$$

where $pF(t)$ is the personal-fitness value at time t, $\boldsymbol{X}(t)$ is the particle's position, $\boldsymbol{pT}$ stores the color the particle is looking for and $I(\boldsymbol{X}(t))$ represents the color of the pixel at the coordinate $\boldsymbol{X}(t)$. Finally, $d_{YUV}(\boldsymbol{a}, \boldsymbol{b})$ represents a distance-like measure in the YUV colorspace:

$$d_{YUV}(\boldsymbol{a}, \boldsymbol{b}) = \begin{cases} \frac{|a_u - b_u| + |a_v - b_v|}{2.0} & \text{if } |a_u - b_u| < T \wedge |a_v - b_v| < T \\ +\infty & \text{otherwise} \end{cases} \tag{2}$$

where T is a pre-set color threshold. Hence the personal fitness has values in the range $[0, 1]$.

The second term aims at estimating how regularly the particles are distributed. To avoid doing so by computing some global statistic of the whole swarm, an effective simple index which can be defined is the variance of the neighbors' distances: when a particle is equally distant from its neighbors, this small "sub-swarm" is well structured and the variance of these distances is zero. Otherwise the variance increases with the increment of disorder in the particles' relative positions.

$$oF(t) = 1.5 \cdot \left(1.0 - \frac{1.0}{1.0 + e^{5.0 - \sigma_d}}\right) \tag{3}$$

where σ_d is the variance of the neighbors' distances. This second term can add up to 1.5 points to the total fitness.

The third term aims at evaluating the "fitness" of the particles' neighbors.

$$nF(t) = 2.5 \cdot \frac{\sum_{i=1}^{nn} tF_i(t-1)}{5.0 \cdot nn} \tag{4}$$

It is defined as the average total fitness of the nn neighbors of the particle, re-scaled in the interval $[0, 2.5]$ (the maximum value tF can assume is 5.0). Therefore, this is the term which has the highest weight in computing one particles' "total fitness".

2.3. Modified Velocity Equation for PM-PSO

The velocity update equation we found to be effective is the following:

$$V(t) = \chi(pF) \cdot \left\{ \begin{array}{l} C_1 \cdot rand() \cdot [X_{best}(t-1) - X(t-1)] \\ + C_2 \cdot rand() \cdot [X_{nb}(t-1) - X(t-1)] \end{array} \right\} \tag{5}$$

where pF is the personal fitness of the particle, $\chi(pF)$ is the constriction factor (expressed as a function of pF), $X_{best}(t-1)$ is the best position recently visited by the particle, $X_{nb}(t-1)$ is the neighborhood barycenter.

The only problem which arises when using the barycenter as an attractor is observable on the borders of the lattice: since individuals in those positions have less than four neighbors, they tend to be attracted towards the group's barycenter, causing the swarm to "shrink" within an area which is smaller than expected. To overcome this problem, as well as to let the external part of the grid fit the template's limits better, we added a repulsion term in the velocity update equation for all those particles who have less than four neighbors:

$$rep(t) = -C_3 \frac{gB(t-1) - X(t-1)}{\|gB(t-1) - X(t-1)\|} \cdot e^{tF(t-1)-5.0} \tag{6}$$

where C_3 is a repulsion factor and gB represents the group's barycenter. This term is not affected by the constriction factor: in this way the lattice borders tend to stretch the grid when their total fitness increases. This grid widening stops when the swarm has completely covered the template, even if its external particles keep jumping in and out of the template.

The next subsections describe all terms in Eq. (5) in details.

2.3.1. *Constriction Factor* χ

When a particle finds a color which is very similar to its target, it is worth slowing down its velocity to better analyze that area and wait for its neighbors to come and bring help. It has been observed that, without slowing down, the system is less effective. The function we eventually found to be suitable for this purpose is the following:

$$\chi(pF) = \begin{cases} 1.0 & \text{if } pF <= 0.5 \\ 0.3 + (1.0 - pF) \cdot 1.4 & \text{if } pF > 0.5 \end{cases} \tag{7}$$

It represents a linear smoothing from 1.0 down to 0.3 for values of pF greater than 0.5.

2.3.2. *Personal Best Position*

Since PM-PSO is designed to work also in a dynamic environment, we adopted the "forgetful particles" paradigm.[10] In this way, the personal best position does not coincide with the best position visited by the particle so far, by it is just the best position visited during the last mS (memorySize) steps.

$$\boldsymbol{X}_{best}(t-1) = \boldsymbol{X}(t-n), \quad \text{where } n : pF(t-n) = \max_{i=1,mS} pF(t-i) \tag{8}$$

Obviously, mS becomes another parameter of the system which needs to be correctly tuned for the swarm to achieve a good dynamic behaviour.

During this investigation we had to decide whether to base the selection of $\boldsymbol{X}_{best}$ on a particle's personal fitness pF or on total fitness tF. Even if it might seem more logical to adopt tF as a criterion, the system showed to be more robust to noisy images using pF as described above in Eq. (8).

2.3.3. *Neighborhood Barycenter*

As expressed by Eq. (5), the social attraction factor is set to be the neighborhood barycenter. This is necessary, and logical, to let the swarm reorganize according to the same grid structure it had when it was placed on the template during initialization. This reorganization should take place only when the total fitness of the particles starts to increase. At the same time, a single particle should be attracted only by those neighbors who have high fitness, to avoid being distracted by the others. Consequently, we compute the neighborhood barycenter as follows:

$$\boldsymbol{X}_{nb}(t-1) = \frac{\sum_{i=1}^{nn} \boldsymbol{X}_{best_i}(t-1) \cdot pbF_i(t-1)}{\sum_{i=1}^{nn} pbF_i(t-1)} \tag{9}$$

where the sums are extended to the nn neighbors, $\boldsymbol{X}_{best_i}$ is the personal-best position and pbF_i is the personal-best fitness of neighbor i. Therefore, the barycenter is defined as the mean of the best positions remembered by the neighbors, weighted by their personal best fitnesses. This way the barycenter position is biased towards the fittest neighbors.

Regarding the denominator in Eq. (9), it may happen that all neighbors have a null fitness value, especially in the initial searching phase. This is mostly due to the high neutrality of the fitness landscape seen by each particle. To avoid having problems with divisions by zero, we set $\boldsymbol{X}_{nb}$ equal to the geometric barycenter when the denominator of Eq. (9) is zero.

3. Results

The code for the system is written in C++. Experiments were run on a PC equipped with an Intel® Core™2 Duo processor running at 3.80GHz, letting the swarm search the images for half a second while collecting statistics on the particles' behaviour. Each image was analyzed ten times to verify the repeatability of the experiments: the system appeared to be characterized by a very high repeatability since no major bias in its behaviour could be observed.

The parameters used in the tests are: a grid of 15×15 particles, $C1 = 1.5$, $C2 = 2.3$ and $C3 = 2.0$. Furthermore, the threshold T in the YUV space for the personal fitness was set to 10 and the particles' memory was limited to the last five positions visited ($mS = 5$).

Fig. 1.   The template pattern utilized in the first experiment.

Figure 1 shows the pattern to be matched in the first set of tests performed on synthetic 640×480 images. As a first step, we wanted to check the invariance of the swarm behaviour to basic transformations of the pattern. We tested our system against rotation, scaling, mirroring, skewing and perspective transforms. In Fig. 2 some of these results are shown. The last three outputs refer to harder situations: in the first one the white background was substituted by a pattern containing colors similar to the template to be detected, while in the last two random noise was added to the background and to the whole image, respectively. As can be observed from these images, the swarm correctly self-reorganizes with its underlying lattice structure matching the template. Except for the case with perspective deformation and the one in which noise was added also to the pattern, the system appears able to both detect and precisely locate its target: a bounding-box could easily be depicted by connecting the four particles at the corners of the lattice.

The next set of tests aimed at verifying the ability of PM-PSO to distinguish among very similar patterns. Figure 3 shows some results obtained when just one fake pattern appears in the image to be analyzed and when one fake pattern is there along with the one to be detected. As could be expected, the swarm is unable to self-organize on the fake pattern, while most frequently it could locate the correct pattern when both patterns were present (see the last case in Fig. 3).

Finally, we assessed whether it was possible to understand when the swarm has converged on the right template. To do so we plotted two different indices at each generation: the average total fitness of the swarm and the number of moving particles (those which changed their position in the last step). As can be seen from Fig. 4, they seem to be good indicators of the status of the swarm.

In the case reported on the left, corresponding to the rotated pattern, the mean fitness value quickly reaches values close to five (the maximum possible value), while the number of moving particles rapidly decreases to about fifty. If we recall that the particles at the lattice borders are always kept moving by the repulsion from the swarm barycenter, and that in a 15×15 grid these particles are fifty-six, this could mean that almost all particles have converged to a stable position. In the case on the right, this does not happen: even at the end of the run, about half the particles are still moving and the mean fitness value is largely below five.

Regarding convergence speed, in almost all tests run on images containing the pattern to be detected, we observed that the system needed about $200 \div 300$ generations to converge on a 640×480-pixel image, regardless of the transformation applied to the pattern. Since in half a second our system is able to perform about 2000 position updates for the whole swarm, the mean convergence time can be estimated to be between 50 ms and 75 ms. It is worth mentioning that, on the same system, the matching method

Input Image	Output Detail	Notes

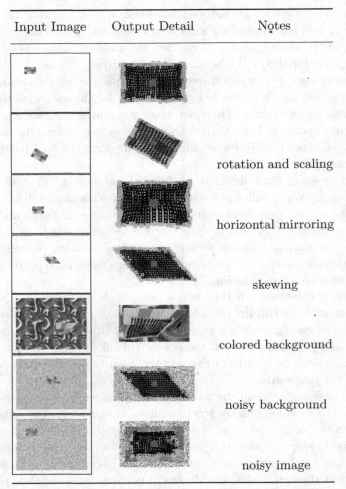

rotation and scaling

horizontal mirroring

skewing

colored background

noisy background

noisy image

Fig. 2. Sample results from the first set of tests. In the central column it is possible to see a magnified detail of a typical output image for the corresponding inputs.

provided with the free version of the well known openCV[a] library needs 270 ms on average to match the same pattern used in our test, in the absence of any kind of deformation. Hence, PM-PSO showed to be faster and more powerful compared to a standard pattern matching technique based on window correlation.

---

[a]http://www.intel.com/technology/computing/opencv/index.htm

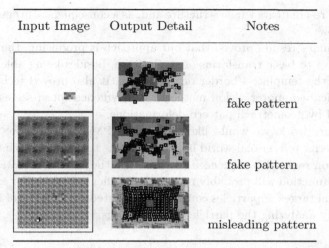

Input Image	Output Detail	Notes
		fake pattern
		fake pattern
		misleading pattern

Fig. 3. Sample results from the second set of test. In the central column it is possible to see a magnified detail of a typical output image for the corresponding input.

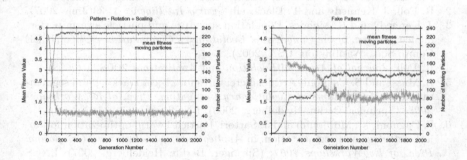

Fig. 4. Trend of the two indices to be possibly used to detect swarm convergence: on the left a typical graph showing good convergence and detection; on the right, results obtained with a fake, even if very similar, pattern.

4. Conclusions and Future Directions

We have presented a PSO variant, called PM-PSO, which can match templates on digital images. In this variant, particles are organized according to the Von Neumann topology and each one has a sample of the pattern as target. In brief, having substituted the social attraction term in the standard velocity equation with an attraction term towards the neighborhood barycenter permitted the swarm to self-reorganize its underlying lattice over the pattern. When the template is not present in the image, the swarm is not able to re-create its whole structure and, as a consequence, fitness values are very low.

Preliminary results proved that our approach is promising: the system is insensitive to basic transforms of the pattern, besides being able to precisely find the template's border on the image. It also proved to be faster than classical computer vision methods. Convergence speed seems not to be affected by geometrical pattern deformations.

As future work, we would like to test our PM-PSO on more complicated patterns and on real-world images, to make this system able to track templates on real video sequences. To this aim, the velocity equations and the fitness function will probably need to be further modified. For example, the personal target of particles could be substituted by a vector of features obtained by analyzing the particle's surroundings.

References

1. J. Kennedy and R. Eberhart, Particle swarm optimization, in *Proc. IEEE Int. conf. on Neural Networks*, (IEEE Comp. Soc., Washington, USA, 1995).
2. R. Poli, J. Kennedy and T. Blackwell, *Swarm Intelligence* 1, 33(June 2007).
3. J. Kennedy and R. Mendes, Population structure and particle swarm performance, in *Proc. of the Congress on Evolutionary Computation - CEC*, (IEEE Comp. Soc., Washington, USA, 2002).
4. R. Poli, *Journal of Artificial Evolution and Applications* **2008**, 1 (2008).
5. L. Mussi and S. Cagnoni, Artificial creatures for object tracking and segmentation, in *Applications of Evolutionary Computing: Proc. of EvoWorkshops 2008*, (Springer, Berlin, Heidelberg, 2008). LNCS vol. 4974.
6. S. Cagnoni, M. Mordonini and J. Sartori, Particle swarm optimization for object detection and segmentation, in *Applications of Evolutionary Computing: Proc. of EvoWorkshops 2007*, (Springer, Berlin, Heidelberg, 2007). LNCS vol. 4448.
7. L. Bocchi, *Evolution of an Abstract Image Representation by a Population of Feature Detectors*, in *Genetic and Evolutionary Computation for Image Processing and Analysis*, eds. S. Cagnoni, E. Lutton and G. Olague, EURASIP - Signal Processing and Communications, Vol. 8 (Hindawi Publishing Corp., New York, USA, 2008), New York, USA, pp. 157–175.
8. L. Anton-Canalis, M. Hernandez-Tejera and E. Sanchez-Nielsen, Particle swarms as video sequence inhabitants for object tracking in computer vision, in *Proc. ISDA '06*, (IEEE Comp. Soc., Washington, USA, 2006).
9. J. Louchet, *Genetic Programming and Evolvable Machines* 2, 101(June 2001).
10. C. Di Chio and P. Di Chio, EcoPS - a model of group-foraging with particle swarm systems, in *Proc. of the 9th European Conference on Advances in Artificial Life - ECAL*, (Springer, Berlin, Heidelberg, 2007). LNCS vol. 4648.

"ADMISSIBLE METHOD FOR IMPROVED GENETIC SEARCH IN CELLULAR AUTOMATA MODEL (AMMISCA)": A STRATEGY IN GENETIC CALIBRATION - PRELIMINARY RESULTS

R. UMETON* and S. DI GREGORIO

Department of Mathematics, University of Calabria
Arcavacata di Rende, CS 87036, Italy.
www.unical.it

Genetic Algorithms (GAs) are widely used to incrementally reach admissible solutions for hard problems such as parameter tuning in Cellular Automata (CA) models. This paper presents a genetic strategy, specifically developed for CA model calibration, exploiting the circumstance that the considered CA parameters have a physical meaning. The proposed approach has proved to be comparable and, in some cases outperforming, if compared with the standard GA proposed by Holland. As a further result, the goodness of the proposed genetic strategy opens the door to genetic tuning algorithms lacking of a standard crossover operator. Results, though preliminary, can be considered encouraging and routes to a wider analysis of the proposed approach.

Keywords: genetic algorithms; model calibration; cellular automata; lava flows.

1. Introduction

In the field of risk assessment and hazard mitigation, event simulation and predictor models have acquired a relevant position. In fact, through simulation of reliable models, risks associated to such processes can be evaluated and possibly predicted and contrastated.

Cellular Automata[1] (CA) proved[2-5] to be a valid choice in simulating natural phenomena such as landslides, erosion processes, lava and pyroclastic flows. They are parallel computing models, discrete in space and time, whose dynamics is determined by the application of local rules of evolution defining the CA transition function. In particular, above cited examples

---

*Corresponding author. Tel.: +39 0984 496467; fax: +39 0984 496410.
Email addresses: umeton@mat.unical.it (R. Umeton), dig@unical.it (S. Di Gregorio).

are based on the Di Gregorio and Serra's approach[6] for the modelling of spatially extended dynamical systems. Models based on this approach generally depend on many parameters, which must be provided with the highest possible accuracy in order to obtain satisfactory results in simulating the considered phenomenon. To do this, a parameter tuning phase through standard GA has been successfully applied in previous works[7–10] .

Genetic Algorithms (GAs)[11,12] are parallel, general-purpose, search algorithms inspired by Genetics and Natural Selection. They simulate the evolution of a population of candidate solutions of a specific search problem by favouring the "survival" and the "recombination" of the best ones, in order to obtain better and better solutions. These family of algorithms have acquired an important role in all those fields dealing with intrinsically-hard problem lacking of dedicated heuristics or ad-hoc algorithms.

This paper proposes the definition of AMMISCA, a genetic strategy, and its application to the parameter tuning of the SCIARA-R7[13] CA model for lava flow simulation and forecasting. The next section presents the SCIARA-R7 simulation model. Section 3 details the AMMISCA genetic strategy, while the fourth section discusses obtained results. Conclusions are reported at the end of the paper.

2. The SCIARA-R7 model

The physical behaviour of lava flows can be partially described in terms of Navier-Stokes equations. Analytical solutions of these differential equations are a hopeless challenge, except for few simple, not realistic, cases. The complexity of the problem resides both in the difficulty of managing irregular ground topography and in complications of the equations, that must also be able to account for flows, exhibiting a wide diversity in their fluid-dynamical behaviour due to cooling processes. An alternative approach to PDE numerical methods for Navier-Stokes[14] (or more complex) equations is offered by Cellular Automata (CA). As announced above, they are computational models assuming discrete space/time and easily implementable on parallel computers. CA SCIARA-R7 for lava flows is derived from SCIARA[4] where the space is a plane, divided in hexagonal cells; each cell is characterised by a state, that specifies the mean values of physical quantities in the cell (e.g. substate altitude) and embodies a computing unit. This unit updates synchronously the substate values according to a transition function on the basis of substate values of the cell and its adjacent ones. The transition function is applied by the sequential computation of "elementary processes", that account for the phenomenon features.

From a formal point of view SCIARA-R7 is stated by the septuple
$SCIARA\text{-}R7 = \langle R, L, X, S, P, \sigma, \gamma \rangle$, where

- $R = \{(x, y) | x, y \in \mathbb{N}, 0 < x < l_x, 0 < y < l_y\}$ is the set of identical hexagonal cells identified by integer co-ordinates in the finite region where the phenomenon evolves.
- $L \in R$ specifies the lava source cells (i.e. craters).
- X identifies the geometrical pattern of cells that influence the cell state change. They are, respectively, the cell itself and its adjacent cells: $X = \{(0,0), (0,1), (0,-1), (1,0), (-1,0), (-1,1), (1,-1)\}$.
- $S = Q_A \times Q_{th} \times Q_T \times Q_O{}^6$ is the set of states; more in detail, Q_A is the altitude of the cell, Q_{th} is the thickness of lava inside the cell, Q_T is the lava temperature and $Q_O{}^6$ rappresent lava outflows (6) from the central cell towards the adjacent ones.
- $P = \{p_{clock}, p_{TV}, p_{TS}, p_{chlV}, p_{chlS}, p_{adher}, p_{cool}\}$ is the set of global parameters, in which:

  - p_{clock} is the time corresponding to a CA step
  - p_{TV} is the lava temperature at vent
  - p_{TS} is the lava solidification temperature
  - p_{chlV} is the characteristic length at the vent temperature
  - p_{chlS} is the characteristic length at the solidification temperature
  - p_{adher} is the constant adherence of lava passing on a cell
  - p_{cool} is the cooling parameter

- $\sigma : Q^{6+1} \rightarrow Q$ is the deterministic state transition function, which is simultaneously applied to all cells of the CA.
- $\gamma : Q_{th} \rightarrow \mathbb{N} \times Q_{th}$ specifies the emitted lava from source cells at the CA step $t \in \mathbb{N}$.

In order to evaluate the goodness of simulations obtained with the detailed model, we have adopted the evaluation function $e_2 = \sqrt{\frac{R \cap S}{R \cup S}}$ where R and S represent the area covered by simulated and real lava flow, respectively; this evaluation function is then used to compute the fitness associated to each simulation in the genetic process.

The next Section describes the proposed genetic strategy for optimizing SCIARA-R7 parameters by means of the AMMISCA approach.

3. AMMISCA in detail

AMMISCA, the acronym of AdMissible Method for Improved genetic Search in Cellular Automata, is a genetic strategy exploiting the circumstance that each element of the set of parameter to be tuned (P) has a physical meaning. For instance, if $parent_A$ and $parent_B$ expresses the p_{chlV} SCIARA-R7 parameter (which represents a "threshold" for lava mobility) with values 15 and 25 meters respectively, it can be erroneous to assign the next offspring to an improbable value of 50 meters (which is too *distant* from the parent contributes). As anticipated above, the main, characterizing, difference between a standard Holland GA and AMMISCA regards the field which they have been designed for (Cfr. Fig. 1). While the standard GA is a general purpose optimizer, the second one has been designed for the resolution of those problems in which parameters encoded in the individual have a physical correspondence. When this physical correspondence exists, our algorithm takes advantage of it, thanks to the different crossover strategy implemented, which strictly preserves previous obtained results.

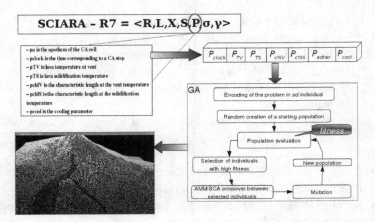

Fig. 1. The tuning process: (1) select the part of the model that has to be tuned: the parameter set in our case; (2) encode this parameter set in the individual; (3) run the Genetic Algorithm in order to let admissible solutions evolve and recombine, favouring better solutions: in our case the fitness is evaluated through the function e_2; (4) extract the parameter set that gave the most realistic simulation; (5) adopt this set to complete the lava forecasting model.

The basic idea within the AMMISCA strategy is to go beyond the preservation of promising schemes through a different crossover, based on arithmetic average: while a one-point crossover ($ONEPT$), using a randomly selected crosspoint, can transform parent strings (e.g. AAAAA, BBBBB) into quite different strings (e.g. AABBB, BBAAA), our crossover calculates for each parameter the average value between parent ones (as proposed in Linear crossover[15] method with *weight* = 0.5), and assigns it to the next generation allele. From two parents we get only one offspring; moreover, this single individual might be too much specialized and the average-driven recombination seems to converge too much rapidly. In order to solve these problems, a sort of *anti-dimidium* is introduced. In AM-MISCA, as in standard GAs, we have a range for each parameter encoded in the individual, and two points inside the range rappresenting the value of the parameter introduced by parents. If we shift from a linear range to a closed one (Cfr. Fig. 2), we obtain a circumference where minimum and maximum of the range coincide and, while the average value is assigned to the first offspring (i.e. $P_{A_{i+1}} = (P_{A_i} + P_{B_i})/2$) as the logical middle-point between parent values, the *anti-dimidium* is calculated as the point diametrically opposite to it (i.e. $P_{B_{i+1}} = P_{A_{i+1}} + (P_{max} + P_{min})/2$).

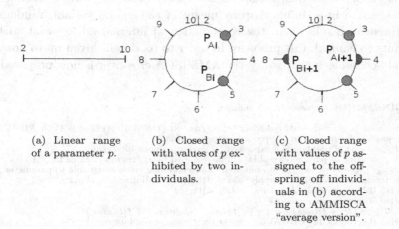

(a) Linear range of a parameter p.

(b) Closed range with values of p exhibited by two individuals.

(c) Closed range with values of p assigned to the off-spring off individuals in (b) according to AMMISCA "average version".

Fig. 2.   The shift from the linear range to the closed one along with average and *anti-dimidium* definition.

An idea, subtended by the introduction of *anti-dimidium*, concerns the following problem: some couples of parameters in SCIARA are "antagonist". This means that similar results could be obtained increasing the value of the former parameter and decreasing the value of the latter one. Hence, different clusters of good values of parameters may exist. Then, AMMISCA always proposes an "internal" ($P_{A_{i+1}}$ in Fig. 2) allele and an "external" ($P_{B_{i+1}}$) one: the former searches for a specialization of parents, while the latter explores values out of the interval defined by parents. Finally, in the context of SCIARA clustered parameters, AMMISCA conveniently derives a new offspring by composing "internal" and "external" alleles (Cfr. last line of following pseudo-code block). Besides the application of the mere average for model calibration as described above, this work presents two further variants of the algorithm which consider different offspring calculations. In particular, the first version uses the fitness associated to every parent in order to weigh their contribution and thus is labeled as a "fitness weighted average" ($FWAVG$); indeed, the more a parent is promising, the *closer* the allele will be to it. The second variant chooses a random point inside the sub-interval delimited by parents (denoted as $RWAVG$ and proposed in Heuristic crossover[16]). The pure application of the versions detailed above could result too fitness-driven and interfer with crossover function and research space inspection (the first variant, fitness weighted average), or could take longer to solve easy problems (e.g. a maximum values search in a simple cusp by means of the second variant, randomly weighted average). Then, the combination of internal ad external alleles permits to embank this problem. In order to fix details given up to now, a pseudo-code-block that states the AMMISCA crossover is now proposed.

```
BEGIN: AMMISCA_crossover_function()
{
    crossmode = get requested crossover type //one in {ONEPT, AVG, FWAVG, RWAVG}
    for each (parameter p in P encoded in the individual)
    P_A = value of parameter p expressed by parent_A. Same for P_B.
    P_A' = value of parameter p that will be assigned to offspring_A. Same for P_B'.
    range_min and range_max are minimum and maximum value assignable to parameter p
    if (crossmode==ONEPT) applyStandardCrossoverByHolland(P_A,P_B,P_A',P_B');
    else if (crossmode==AVG) P_A' = (P_A + P_B)/2;
    else if (crossmode==FWAVG)
        P_A' = (P_A * fitness_A + P_B * fitness_B)/(fitness_A + fitness_B);
    else if (crossmode==RWAVG) //uses only positive random numbers
        P_A' = (P_A * random_1 + P_B * random_2)/(random_1 + random_2);
    if (P_A' == (range_min + range_max)/2)
        P_B' = choose randomly, with same probability, between range_min and range_min
    else P_B' = P_A'+ half round of the range; //anti-dimidium
    if (P_B' > range_max) P_B' = P_B' - range_max + range_min
    if (crossmode ≠ ONEPT) swap P_A' and P_B' with proability 0.5; //alleles composition
} END;
```

In the next section we briefly present the main results achieved by AMMISCA applied to the calibration of the model SCIARA-R7.

4. AMMISCA results

In order to validate the genetic strategy for a parameter tuning task, AM-MISCA is used for the calibration of SCIARA-R7 model applied to the Nicolosi lava flow event which occurred at Mt Etna (Italy) in 2001.

We consider different classes of tests: first, we use many seeds and few GA generations (50 seeds and 10 generations, Cfr. Fig. 3 for setup details), and subsequently we adopt the most promising seeds for further GA generation computation (i.e. the most promising seed for 100 generations). In Table 1, the first test as a result of about 21000 individual evaluations is presented, where the four adopted algorithms (single-point crossover and AMMISCA in its three versions) are compared by means of the contribution of the best average in the individual pool and the best individual. The most promising seed of each algorithm is further executed for ninety more generations and Fig. 4 displays both fitness trend and time comparison.

Numbers of seeds	50

GA Setup	
Parameters num.	7
Initial number of individuals	16
Individuals replaced at each step	8
Crossover probability	1
Mutation probability	0,083
GA steps	10

Individual composition			
Parameter (unit)	NO. of bits	Range-min	Range-max
clock (s)	8	60	240
TV (K)	0	1323	1323
TS (K)	8	1023	1173
chIV (m)	4	0,1	5
chIS (m)	4	5,1	10
adher (m)	8	0,05	10
cool (m³/K³)	8	10^{-19}	10^{-12}

Fig. 3. The GA setup for the first class of test.

As a result, the AMMISCA strategy proves to be valid and promising, being able to outperform standard GA with single-point crossover, both in terms of obtained fitness and execution times. Table 1 and Fig. 4 indicate that AMMISCA obtains the best individual in both test cases (10 and 100 GA iterations), giving thus the most precise lava event simulation. Besides these results, AMMISCA chooses a set of individuals characterized by a

106

Table 1. First test set: generation-by-generation, for 10 generations, which algorithm gives the best results, over 50 seeds evaluation for each algorithm.

Generation	Best average fitness	Best individual
1	AVG	ONEPT
2	AVG	ONEPT
3	AVG	FWAVG
4	ONEPT	FWAVG
5	ONEPT	FWAVG
6	ONEPT	FWAVG
7	ONEPT	FWAVG
8	ONEPT	AVG
9	ONEPT	AVG
10	**ONEPT**	**AVG**

Note: ONEPT is one-point crossover; AVG is AM-MISCA average version; FWAVG is fitness weighted average version; RWAVG is randomly weighted average version.

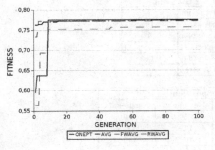

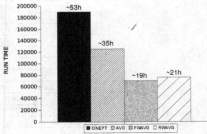

(a) Fitness trend evolution of each algorithm over 100 GA generations as emerged in second test set: keep running the most promising (at 10th generation) seed for each algorithm for 90 generations more.

(b) Total time required by each algorithm over the two test sets, remarking the fact that some of them explored search zones ignored by others, at least with respect to p_{clock} parameter.

Fig. 4. Second test set results and global time required by each algorithm to complete the two tests.

high P_{clock} value leading to faster computations and lower execution times; such P_{clock} were not taken into account by the Holland search strategy.

5. Conclusions and future developments

Results can certainly be considered encouraging for this initial stage in the study of AMMISCA genetic strategy. Moreover, besides the fact that AMMISCA gives rise to the most precise lava simulation, it is interesting to note that we achieve the best solution (in terms of fitness and required time) without a standard crossover phase as defined by Holland. Furthermore, these results route to ad-hoc tuning techniques for CA models that are similar to the analyzed one, that are CA models where the parameter set has a physical meaning.

As a consequence of these preliminary results, AMMISCA can be more deeply inspected as an alternative to a standard GA algorithm. The game plan for future work is to study the AMMISCA conduct in the calibration of SCIARA model for factitious lava events[17] (the best simulation is considered as the real lava event). In fact, we can better compare standard GAs and our family of algorithms with respect to an artificial lava event so that theoretically the global optimum can be achieved during calibration. To be more precise, the referring artificial simulation can be either the simulated lava event obtained with Holland's GA or the one obtained with AMMISCA "average version" (respectively the first and the second simulation whose fitness is rappresented in Fig. 4). Subsequently, the second step in this validation plan is to use our family of algorithms to calibrate other macroscopic CA models, tuned with standard GA in the past, such as SCIDDICA[2], PYR[5] and SCAVATU[3]. Eventually, through the analysis of AMMISCA behaviour on cited models, we can derive a study of fitness landscape[17] and reach a more accurate idea of the AMMISCA convergence process.

Acknowledgments

Authors thank Dott.ssa Maria Vittoria Avolio, Dott. William Spataro, Dott. Rongo Rocco and Dott. Donato D'Ambrosio for their support during the development of the work and for their advices in the paper redaction.

References

1. J. von Neumann, *Theory of Self Reproducing Automata* (University of Illinois Press, 1966).
2. S. Di Gregorio, R. Rongo, C. Siciliano, M. Sorriso-Valvo and W. Spataro, *Phys. Chem. Earth Pt. A* **24**, 97 (1999).
3. D. D'Ambrosio, S. Di Gregorio, S. Gabriele and R. Gaudio, *Phys. Chem. Earth Pt. B* **26**, 33 (2001).

4. G. M. Crisci, S. Di Gregorio, R. Rongo and W. Spataro, *J. Vol. Geo. Res.* **132**, 253 (2004).
5. G. M. Crisci, S. Di Gregorio, R. Rongo and W. Spataro, *Fgcs* **21**, 1019 (2005).
6. S. Di Gregorio and R. Serra, *Fgcs* **16**, 259 (1999).
7. D. D'Ambrosio, S. Di Gregorio and G. Iovine, *Nhess* **3**, 545 (2003).
8. P. M. Atkinson, G. M. Foody, S. Darby and F. Wu, *Brains versus Brawn - Comparative Strategies for the Calibration of a Cellular Automata-based urban growth model*, in *GeoDynamics*, (CRC Press, 2004).
9. B. Straatman, R. White and G. Engelen, *Comput. Environ. Urban* **28**, 149 (2004).
10. D. D'Ambrosio and W. Spataro, *Parallel Comp.* **33**, 186 (2007).
11. J. H. Holland, *Nonlinear environments permitting effcient adaptation*, in *Computer and Information Sciences II*, (New York: Academic, 1967).
12. J. H. Holland, *Adaptation in Natural and Articial Systems* (University of Michigan Press, Ann Arbor, 1975).
13. M. Avolio, G. M. Crisci, S. Di Gregorio, R. Rongo and R. Umeton, Introduction of more physical features in the cellular automata model for lava flows sciara: preliminary results regarding the viscosity (2007), Asia and Oceania Geosciences Society.
14. H. Lamb, *Hydrodynamics* (Cambridge University Press, 1879).
15. A. H. Wright, Genetic algorithms for real parameter optimization, in *Foundations of genetic algorithms*, ed. G. J. Rawlins (Morgan Kaufmann, San Mateo, CA, 1991) pp. 205–218.
16. A. H. Wright, Genetic algorithms for real parameter optimization, in *Foundations of Genetic Algorithms, First Workshop on the Foundations of Genetic Algorithms and Classier Systems*, ed. G. J. Rawlins (Morgan Kaufmann, San Mateo, CA, 1990) pp. 205–218.
17. D. D'Ambrosio, W. Spataro and G. Iovine, *Comput. Geosci.* **32**, 861 (2006).

PART III

COGNITION

EVOLVING NEURAL WORD SENSE DISAMBIGUATION CLASSIFIERS WITH A LETTER-COUNT DISTRIBUTED ENCODING

A. AZZINI and C. DA COSTA PEREIRA and M. DRAGONI and
A. G. B. TETTAMANZI

Università degli Studi di Milano
Dipartimento di Tecnologie dell'Informazione
Via Bramante 65, I-26013 Crema (CR), Italy
E-mail: azzini,dragoni,pereira,tettamanzi@dti.unimi.it

We propose a supervised approach to word sense disambiguation (WSD), based on neural networks combined with evolutionary algorithms. Large tagged datasets for every sense of a polysemous word are considered, and used to evolve an optimized neural network that correctly disambiguates the sense of the given word considering the context in which it occurs. The viability of the approach has been demonstrated through experiments carried out on a representative set of polysemous words.

Keywords: evolutionary algorithms; artificial neural networks; supervised learning; word sense disambiguation.

1. Introduction

The automatic disambiguation of word senses consists of assigning the most appropriate meaning to a polysemous word, i.e., a word with more than one meaning, within a given context. This consists of two steps: (i) considering the possible senses of the given word; and (ii) assigning each occurrence of the word to its *appropriate* sense.

We propose a supervised approach to word sense disambiguation based on neural networks (NNs) combined with evolutionary algorithms (EAs). We dispose of large tagged datasets describing the contexts in which every sense of a polysemous word occurs, and use them to evolve an optimized NN that correctly disambiguates the sense of a word given its context. To this aim, we use an EA to automatically design NNs, with a distributed encoding scheme, based on the way words are written, to represent the context in which a word occurs. The paper is organized as follows: Section 2 briefly describes the EA used in this work, whereas Section 3 explains its application to WSD, followed, in Section 4, by the results of experiments. Section 5 concludes with a discussion of the results.

2. The Neuro Evolutionary Algorithm

Their tolerance for noise, their ability to generalize, and their well-known suitability for classification tasks[1] make NNs natural candidates for approaching WSD. A large number of successful applications demonstrates that NN design is improved by considering it in conjunction with EAs,[15] and one of the most promising approaches to using EAs to design and optimize NNs jointly evolves network architecture and weights, without any intervention by an expert.[1]

We use an evolutionary approach for NN design, previously validated on different benchmarks and real-world problems,[3,4] which uses a variable-size representation of individuals. A population of classifiers — the individuals — is maintained, encoding multilayer perceptrons (MLPs), a type of feed-forward NN. The evolutionary process is based on the joint optimization of NN structure and weights, and takes advantage of the error back-propagation (BP) algorithm to decode a *genotype* into a *phenotype* NN. Accordingly, it is the genotype which undergoes the genetic operators and which reproduces itself, whereas the phenotype is used *only* for calculating the genotype's fitness. The rationale for this choice is that the alternative of using BP, applied to the genotype, as a kind of "intelligent" mutation operator would boost exploitation while impairing exploration, thus making the algorithm too prone to being trapped in local optima.

2.1. *Evolutionary Encoding*

Each individual is encoded in a structure in which basic information is maintained as illustrated in Table 1. For each simulation a new population of a specific subset of NN architectures, MLPs, is created. The genotype is represented by an input layer and a number of hidden layers together with the output layer, equal to the layer vector size. The input and output sizes are pre-established at network definition. The input size is 26 neurons (see Section 3). The output size is given by the number of senses of the target word. The number of hidden nodes in the ith hidden layer corresponds to the number specified in the ith element in the topology vector (see Figure 1, top), while the chromosome of the individual is defined by the corresponding MLP (see Figure 1, bottom).

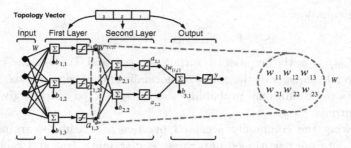

Fig. 1. Genotype representation of the NN: example of a topology vector and its associated weight matrix W.

Individuals are not constrained to a pre-established topology, and the population is initialized with different hidden layer sizes and different numbers of neurons for each individual according to two exponential distributions, in order to maintain diversity among all the individuals in the new population. The number of neurons in each hidden layer is constrained to be greater than or equal to the number of network outputs, in order to avoid hourglass structures, that could reduce the performances. Indeed, a layer with fewer neurons than the outputs destroys information which later cannot be recovered. Initial weights and biases are extracted from a normal distribution. Like in evolution strategies,[5] for each connection weight and neuron bias encoded in the genotype, an associated variance is also encoded, which determines the probability distribution of mutations and is used to self-adapt the mutation step.

Table 1. Individual Representation.

Element	Description
topology	String of integer values that represent the number of neurons in each layer.
$\mathbf{W}^{(0)}$	Weights matrix of the input layer neurons of the network.
$\mathbf{Var}^{(0)}$	Variances for every element of $\mathbf{W}^{(0)}$.
$\mathbf{W}^{(i)}$	Weights matrix for the ith layer, $i = 1, \ldots, l$.
$\mathbf{Var}^{(i)}$	Variances for every element of $\mathbf{W}^{(i)}$, $i = 1, \ldots, l$.
b_{ij}	Bias of the jth neuron in the ith layer.
$\mathrm{Var}(b_{ij})$	Variance of the bias of the jth neuron in the ith layer.

2.2. *Fitness Function*

We adopt the convention that a lower fitness means a better NN, mapping the objective function into an error minimization problem. The fitness of an individual is calculated based on the confusion matrix, by means of the following equation:

$$f = N_{\text{outputs}} - \text{Trace}, \tag{1}$$

where N_{outputs} is the number of output neurons and Trace is the sum of the diagonal elements of the row-wise normalized confusion matrix, which represents the conditional probabilities of the predicted outputs given the actual outputs.

Following the commonly accepted practice of machine learning, the problem data are partitioned into three sets: training, test and validation set, used respectively to train, to stop the training, thus avoiding over-fitting, and to assess the generalization capabilities of a network. In the neuro-genetic approach the fitness is calculated for each individual on the test set after applying BP.

2.3. *Genetic Operators*

The evolutionary process is based on the genetic operators of selection and mutation, described in detail in a previous work,[3] and here briefly reported. Recombination is not used, due to its detrimental effects on NNs.[15]

The selection method implemented in this work uses elitism, allowing the best individual to survive unchanged into the next generation, and solutions to monotonically get better over time. The algorithm uses truncation selection with a proportion of individuals selected for reproduction of $\frac{1}{2}$: starting from a population of n individuals, the worst $\lfloor n/2 \rfloor$ (with respect to f) are eliminated. The remaining individuals are then duplicated in order to replace those eliminated, and finally, the population is randomly permuted.

Two kinds of NN perturbations are used: a general weight mutation, that perturbs the weights of the neurons before performing any structural mutation and applying BP, and a topology mutation, that perturbs the structure maintaining the overall network behavior unchanged.

Weight Mutation defines a Gaussian distribution for the variance matrix values $\mathbf{Var}^{(i)}$ of each weight matrix $\mathbf{W}^{(i)}$, defined in Table 1, much like in *evolution strategies*.[6] All the weight matrices $\mathbf{W}^{(i)}$ and the corresponding biases are perturbed by using variance matrices and evolutionary strategies applied to the synapses of each NN. The main idea behind these strategies is to allow a control parameter, like mutation variance, to self-adapt rather than changing their values by some deterministic algorithm. Weight Control is applied to each NN after weight perturbation. For each hidden layer with a number or neurons greater than the number of outputs, the neurons whose contribution to the network output is negligible (i.e., strongly below-average, as detailed in[3]) are eliminated. Topology Mutation defines then layer addition and elimination and neuron insertion with independent probabilities, corresponding respectively to three algorithm parameters p^+_{layer}, p^-_{layer} and p^+_{neuron}, set at the beginning and maintained unchanged during the entire evolutionary process.

3. Application to Word Sense Disambiguation

The algorithm presented in Section 2 is used to evolve a specific NN specializing in disambiguating one given target polysemous word. The same process must be repeated for each polysemous word one is interested in disambiguating. Since our aim here is to demonstrate the feasibility of this approach, we will focus on a small, but representative, set of target polysemous words.

3.1. *Dataset*

We used IXA Group's web corpus,[2] which comprises all WordNet noun senses. Its construction is inspired by the "monosemous relatives" method.[10] From this web corpus we extracted all data relevant to the words listed in Table 2, with their number of senses. For each of these words 75% of the records have been used for training, 12.5% for testing, and 12.5% for validation.

Two kinds of representations commonly used in connectionism are distributed[9] and localist schemes.[7,11] The major drawback of the latter is the high number of inputs needed to disambiguate a single sentence, because every input node has to be associated to every possible sense; in fact, this number increases staggeringly if one wants to disambiguate an entire text.

Examples of distributed schemes are microfeatures[14] and the word-space model.[13] Distributed schemes are very attractive in that they represent a context as an activation pattern over the input neurons. Therefore, unlike with localist schemes, the number of input neurons required does not have to equal the size of the vocabulary. On the contrary, there is an intrinsic idea of *compression*, whereas each input neuron encodes a given *feature* of a word or sentence and thus can be reused to represent many different words or senses. Of course, the more the input patterns are compressed into a low-dimensional space, the more information is lost. However, despite such information loss, enough information may still be there to allow meaningful processing by the NN. We used a distributed *letter count* representation scheme, which brings the distributed representation idea to one extreme, by using the number of occurrences of the 26 letters of the alphabet as features.

The activation of the input neurons is obtained by summation of the activation patterns representing the words occurring in a given context, excluding the target word, after removing stop words and stemming the remaining words. Additional fields of the training set (one for each output neuron) correspond to the n senses of the target word. They are all set to zero except the one corresponding to the correct sense. For example, starting from a sentence '*part aqueduct system*', where the target word, *tunnel*, has two senses, (1) "a passageway through or under something" and (2) "a hole made by an animal", the associated record would be

$$2\ 0\ 1\ 1\ 2\ 0\ 0\ 0\ 0\ 0\ 0\ 1\ 0\ 0\ 1\ 1\ 1\ 2\ 3\ 2\ 0\ 0\ 0\ 1\ 0 \rightarrow 1\ 0,$$

in which the first 26 numbers represent the occurrences of the letters of the alphabet, and the last 2 the two output senses (here $n = 2$). The rationale for this type of sentence encoding is that, since every word can be regarded as an activation pattern of the input layer of a NN, the overlap of different patterns from the same sentences allows the network to detect significant information about the context.

The success of the experiments presented in Section 4 will serve as an empirical proof that even such extremely compressed representations preserve enough context information to allow disambiguation of word sense.

4. Experiments and Results

Several experiments have been carried out in order to find out the optimal settings of the genetic parameters p_{layer}^+, p_{layer}^-, and p_{neuron}^+. For each run of the EA, up to 250,000 network evaluations (i.e., executions of the network on the whole training set) have been allowed, including those performed by the BP algorithm. The first group of experiments was carried out on the word *memory*, and the best experimental solutions have been found with p_{layer}^+, p_{layer}^- and p_{neuron}^+ equal to 0.05, although they do not differ significantly from other solutions. The best settings found for *memory* are subsequently used for the disambiguation of the remaining words; the results of these experiments are reported in Table 2. The third column in the table provides, as a benchmark, the expected disambiguation accuracy over all words with a given number of senses, as computed by Galley and McKeown,[8] using the semantic concordance corpus (semcor), a corpus extracted from the Brown Corpus and semantically tagged with WordNet

senses. For instance, they have calculated that, on average, WSD methods
have an accuracy of 75% when applied to all words with two senses.

Table 2. Disambiguation Accuracy.

Word	# of Senses	Exp. Acc.	Dataset Size	Max. Dist.	Min. Dist.	Avg. Dist.	Neuro-Gen Accuracy	Mihalcea "Basic" [12]
bar	11	-	59156	23	4	12	33.93	31.45
bum	4	57	25761	11	5	8	42.64	37.20
channel	6	43	34629	17	3	12	32.89	43.18
circuit	7	35	35457	19	5	13	42.16	40.35
church	3	62	9366	17	6	12	46.94	52.77
day	9	35	64072	11	2	5	35.26	45.52
detention	2	75	3971	12	12	12	92.34	79.16
dyke	2	75	5560	13	13	13	97.40	38.61
feeling	7	35	14269	12	3	7	32.06	39.21
grip	5	52	7464	19	10	14	48.93	53.46
hearth	3	62	6856	10	7	8	50.88	44.82
holiday	2	75	17215	6	6	6	88.56	84.61
lady	3	62	20145	7	2	5	46.61	61.53
material	6	43	44962	12	4	8	68.36	37.28
memory	5	52	12963	17	6	10	57.12	—
mouth	8	33	25309	17	3	11	53.26	50.00
post	6	43	21781	20	5	11	59.05	37.93
yew	2	75	11295	17	17	17	75.34	70.00

The results reported in the last column of Table 2 are relevant to the
results obtained by another technique recently reported in the literature,
namely the basic classifiers,[12] that are created by using combinations of
features commonly used in several WSD classifiers.

Out of the 18 words for which comparison data are available, Mihalcea's
basic classifiers outperform our evolved NNs 5 times only, namely for the
words *channel, church, day, feeling,* and *lady.*

It should be noticed that not all senses differ semantically by the same
amount. To this aim, the table reports information about distance among
different senses of each word, measured as the minimum number of ontology
edges connecting two concepts in WordNet. It can be observed that, for
instance, the senses of *day* are quite close to each other. This means that
confusion of senses may be expected, as the contexts in which semantically
close meanings of a word are used may be very similar or even coincide.
As a matter of fact, a quick inspection of the results suggests that better
accuracy is obtained by our method for words whose senses are more apart.

If one factors distance between senses in, the results shown in Table 2
look quite promising. Indeed, for all applications of WSD, what is more
critical is to be able to disambiguate between senses that are semantically
far apart.

118

5. Discussion and Concluding Remarks

To provide a more detailed, critical discussion of the results, we consider the graphical representation of the corresponding row-wise normalized confusion matrices, which may also be regarded as the characteristic matrix C of an information transmission channel. Unfortunately, the results of other related work previously carried out in the literature[12] are not discussed in terms of the confusion matrix, but only of the coarser (and less accurate) percentage of accuracy. The confusion matrices for the 18 words considered with the two representation schemes are shown in Figure 2.

Table 3. Senses of the words *memory* and *holiday*.

Word	Occurrences (%)	Sense
memory	3576 (27.59)	1. Something that is remembered.
	2571 (19.83)	2. The cognitive processes whereby past experience is remembered.
	727 (5.61)	3. The power of retaining and recalling past experience.
	5949 (45.89)	4. An electronic memory device.
	140 (1.08)	5. The area of cognitive psychology that studies memory processes.
holiday	1041 (6.05)	1. Leisure time away from work devoted to rest or pleasure.
	16174 (93.95)	2. A day on which work is suspended by law or custom.

In the representation of the word *memory* the EA classifies with satisfactory performances three of the five senses of the word listed in Table 3, respectively the first, the fourth and the fifth sense. The second and the third senses are not classified as well, in that they tend to be confused with the first sense. A similar observation holds for the words *bar, circuit, mouth, day and grip* as well. This fact is mainly due to the high similarity among some of the word senses, as well as among all the sentences in the dataset associated to each of them. The last two senses of *memory* are, instead, very different from the others, and the evolved classifiers perform quite well in recognizing them, even if the occurrences of the last sense account for but a very small percentage, i.e., 1.08%, of the overall dataset used for word *memory*, as shown in Table 3.

A completely different case is given by considering some words with two senses, like *holiday* and *detention*. Indeed, for these two words the evolved NN uses the "shortcut" of always recognizing only one of the two senses, without considering the other. The unbalance of the dataset for the two senses, visible in Table 3, would seem a possible explanation. However, further experiments with artificially balanced datasets for the same words have shown similar results and, after all, the way fitness is calculated neutralizes the bias of unbalanced data. Anyway, in such cases, the "lazy" strategy of

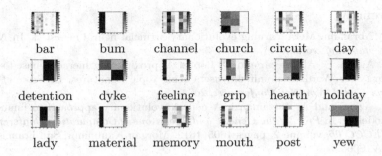

Fig. 2.   Confusion matrices of the words analyzed by the evolutionary approach.

always guessing the most common sense pays off very well.

The performance of "lazy" classifiers would be judged satisfactory with respect to other related works, as indicated in Table 2. Indeed, together with *holiday*, the neuro-genetic approach provides satisfactory results with *detention*, by obtaining an accuracy of 92.34 as opposed to 79.16 obtained from Mihalcea.[12] It should be argued that a "lazy" strategy would be anyway the winning strategy for real-world applications, when what really matters is simply being correct most of the times. Nevertheless, in those cases in which the senses of a word are more far apart, the neuro-genetic classifier obtains satisfactory performances by recognizing all the word senses. An example is given by the words *hearth, yew* and *dyke*, as shown in Figure 2. In particular, for the latter, the system reaches an accuracy of 97.40, as opposed to 38.61 obtained by Mihalcea,[12] which is, by the way, worst than a random choice between the two senses of that word!

At first sight, creating a single NN for every ambiguous word might seem hardly practical or even infeasible. However, there are just 15,935 polysemous words out of the 117,798 WordNet entries. Evolving a NN for disambiguating a polysemous word takes, on an ordinary desktop PC, two hours on average. Assuming some 30 PCs are available day and night, 45 days would be enough to evolve a NN for each polysemous word. We estimate that the entire set of almost 16,000 NNs would occupy 30 Mbytes. When disambiguating a document, a stored NN would be recalled from the database and executed every time a polysemous word were encountered. Recalling a network can take a few milliseconds, whereas executing it is just a matter of microseconds. Therefore, the approach we propose can be considered realistic and feasible with state-of-the-art technology. The next step will be to apply this approach to a larger set of polysemous words and to investigate different input encoding strategies.

References

1. A. Abraham. Meta learning evolutionary artificial neural networks. In *Neurocomputing*, volume 56, pages 1–38, 2004.

2. E. Agirre, E. Alfonseca, and O. Lopez. Approximating hierachy-based similarity for WordNet nominal synsets using topic signatures. In *Proc. of the 2nd Global WordNet Conference.*, 2004.

3. A. Azzini and A. Tettamanzi. A neural evolutionary approach to financial modeling. In *Proc. of the Genetic and Evolutionary Computation Conference, GECCO'06*, volume 2, pages 1605–1612. Morgan Kaufmann, San Francisco, CA, 2006.

4. A. Azzini and A. Tettamanzi. A neural evolutionary classification method for brain-wave analysis. In *Proc. of the European Workshop on Evolutionary Computation in Image Analysis and Signal Processing, EVOIASP'06*, pages 500–504, 2006.

5. T. Bäck, F. Hoffmeister, and H. Schwefel. A survey of evolutionary strategies. In R. Belew and L. Booker, editors, *Prooceedings of the Fourth International Conference on Genetic Algorithms*, pages 2–9, San Mateo, CA, 1991. Morgan Kaufmann.

6. T. Bäck and H. Schwefel. Evolutionary computation: An overview. In *Proceedings of the IEEE International Conference on Evolutionary Computation*, pages 20–29, Nagoya, Japan, May 1996. IEEE Press.

7. G. W. Cottrell. *A Connectionist Approach to Word Sense Disambiguation*. Pitman, London, 1989.

8. M. Galley and K. McKeown. Improving word sense disambiguation in lexical chaining. In *Proc. of the Eighteenth International Joint Conference on Artificial Intelligence*, pages 1486–1488, 2003.

9. G. E. Hinton, J. L. McClelland, and D. E. Rumelhart. Distributed representations. In G. E. Hinton, J. L. McClelland, and D. E. Rumelhart, editors, *Parellel Distributed Processing: explorations in the microstructure of cognition*. MIT Press, Cambridge, MA, 1986.

10. C. Leacock, M. Chodorow, and G. A. Miller. Using corpus statistics and WordNet relations for sense identification. *Computational Linguistics*, 24(1):147–165, 1998.

11. M. Lesk. Automated sense disambiguation using machine-readable dictionaries: How to tell a pine cone from an ice cream cone. In *Proc. SIGDOC Conference*, 1986.

12. R. Mihalcea. Co-training and self-training for word sense disambiguation. In *Proc. of the Conference on Natural Language Learning*, 2004.

13. H. Schütze. Word space. In *Proc. of the 1993 Conference on Advances in Neural Information Processing Systems, NIPS '93*, pages 895–902, San Francisco, CA, 1993. Morgan Kaufmann.

14. D. L. Waltz and Pollack. Massively parallel parsing: A strongly interactive model of natural language interpretation. *Cognitive Science*, 9:51–74, 1985.

15. X. Yao and Y.Liu. A new evolutionary system for evolving artificial neural networks. *IEEE Transactions on Neural Networks*, 8(3):694–713, May 1997.

PREFRONTAL CORTEX AND ACTION SEQUENCES:
A REVIEW ON NEURAL COMPUTATIONAL MODELS*

ILARIA GAUDIELLO

Department of Communication Disciplines, University of Bologna,
Via Azzo Gardino 21, Bologna, 40122, Italy
ilaria.gaudiello@studio.unibo.it

MARCO TULLIO LIUZZA

Department of Psychology, University of Bologna,
Via Berti Pichat, 5, Bologna, 40127, Italy
mtliuzza@gmail.com

DANIELE CALIGIORE

Institute of Cognitive Sciences and Technologies, National Research Council,
(ISTC-CNR)
Via San Martino della Battaglia, 44, 00185, Roma, Italy
daniele.caligiore@istc.cnr.it

The prefrontal cortex (PFC) can be considered the central executive of cognitive control, responsible for the flexibility of human behavior. By a switching-mechanism PFC can update rules and goals representations stored in working memory, so as to performance novel task and accomplish complex routines. PFC functional organization and relation with specific subcortical areas give an account of representations active mantainance that allows to achieve a goal through a series of sub-goals. On the basis of the most recent studies, we present a rieview of the theories concerning PFC role and neural computational models. The paper incudes a focus section on models developed to study the role of PFC in action-sequences learning and performing.

Keywords: Actions sequences; Prefrontal Cortex; Neural networks; Time integration.

*This research was supported by the EU FP7 Project ROSSI, contract no.216125-STREP.

1. Introduction

Human behaviour is characterized by a great flexibility. Flexibility is what allows us to adapt to novel situations and environment contingencies. But this flexibility has a cost: it potentially exposes us to face interferences and confusion. A high order processing of the different stimuli is needed, as well as a temporal organization in order to coordinate huge amount of variable input and consequent beaten in time behaviour. There is a wide agreement on the important role played by PFC in cognitive control. It seems to be involved in flexibility behavioral mechanism and temporal organization,[1,2] internal representations of rules and goals, representations active mantainance,[1] action sequences performance,[3,4] and categorization facilitating higher-order planning based on memory.[5,6]

This paper aim is to review theories and models on the integrative regulation function of PFC as well as its role in temporal organization of action sequences. Hence, the first part of the inquiry consists of a brief description of PFC functional organization, whereas a further inquiry concerns PFC models and its involvement in action-sequences accomplishment. We will focus on two research direction, on the basis of a bio-inspired models[3,7] and a computational framework model.[4,8] A discussion is finally proposed with indication for further research towards biologically plausible models of PFC.

2. The role of PFC and its functional organization

The integrative theory of the PFC function[1] provides a definition of PFC as an active memory in the service of control. In short, this means that PFC is characterized by a robust maintaining of its activity in front of incoming distractions, a multimodal and integrated representations and a high degree of plasticity. Within PFC cerebral functions reaches a high level of integration.

PFC is divided in three regions: orbital, medial an lateral. The first two regions concern the emotional behaviour, while the lateral area is involved in temporal organization of thoughts, actions, language.[2] Moreover, we can give an account of the PFC structure describing it as crossed by two pathways.[9] The controlled pathway includes dorsolateral prefrontal cortex (DLPFC), anterior cingulate cortex (ACC), anterior parts of the cerebellum, anterior automatic pathway includes areas like the supplementary motor area (SMA), primary motor cortex, lateral part of the cerebellum, and lateral part of the basal ganglia.[2] The dorsolateral PFC (DLPFC) - located in the upper side regions of the frontal lobe- and the ventral medial PFC (VMPFC) - located in the innermost regions extending towards the median line and the ventral surface of the frontal lobes. The location of the VMPFC implies close connections with the limbic system.[10] Hence the VMPFC has been implicated both in emotional processing as well as in higher-order sensory processing, it is considered to play an important role in 'decision-making' processes.[11] The location of the DLPFC implies close connections with the sensory and motor areas. Hence, the DLPFC concerns motor control, as well as performance monitoring, goal-directed behavior and executive functions, particularly in the areas of attention and working memory.[12] The DLPFC is also strongly implicated in a task involving the active maintenance and continual updating of recent information. As mentioned before, PFC has a great role in the explanation of flexibility. Flexibility in human behaviour implies the ability of accomplishing habitual task according with rules and goals, and novel task by a mechanism of switching rules. PFC can store in working memory so as to switch the rules of behavior in correspondence to relevant events.

Studies on cerebral development show that the PFC is not completely developed in the early stages of life, and is not completely developed by late adolescence. Rather it is still developing into adulthood, and perhaps achieves maturation only in the third decade of life. A large debate is today focused on whether the DLPFC development occurs in the same amount of time as VMPFC development. Several studies support the idea that the DLPFC has evolved from the motor region and is the later brain region to mature.[13,14] One point is that childhood is characterised by difficulties in performing action-sequences. It has been observed that difficulties in memory enhancement of action-sequences derive from protracted maturation of prefrontal cortex.[15] It has also been showed that children observation of shorter sequences leads to better deferred imitation of single goal actions compared to action sequences.[13,15] Moreover, children ability to identify the goal of an action-sequences is related to their ability to planfully solve a similar sequence. The ability to solve complex problems is close linked to the functions of the working memory, that allows to hold temporarily on-line constraints relevant to the current context. But the development of working memory too, proceeds gradually.

Hence, problems in recall, recognition,performance and encoding of temporal information concerning action-sequences can be attributed to the gradual development of PFC and to the gradual emerging of high-order processing functions, such as active mantainance and goal-directed behaviour.

Flexibility seems to depend on the ability to store abstract rule-like representations. How these representation develop? Instead of rely on representation esplicitly designed for specific tasks, a model should explain the way such representations emerge by: in short, a model should try to simulate a self-organizing system neuroimaging evidence shows that abstract and schematic representations, such as representations of sequential actions, as well as the general rules of motor performances, remain represented in prefrontal networks. The same does not apply for the automatic aspect of motor performances, that can be relegated to lower structure. Further studies give evidence for the coexistence of two neural substrates of active representation: representations for the recent past and representations for anticipated future. The two substrates are anatomical overlapped and belong to the same cortical network of long-term memory.[2]

2.1. *Involvement of PFC on hierarchical organization of behaviour and cross-modality integration*

The high-cognitive processes, as behavioral and linguistic actions, are hierachically organized in the prefrontal cortex, while primary motor and premotor areas constitutes the lowest levels.[16] Koechlin and collegues shows that motor processing and control are processed from anterior prefrontal through caudal prefrontal, to premotor cortex: the information processed from the former level arrives to the next one, moving down in this top-down process.[12] In a task performed by subjects whose were registered their brain ativity by fMRI, Koechlin and collegues showed that stimulus activated premotor area, its context was processed by the caudal prefrontal cortex and the instructional cue by the rostral PFC. These results seems confirm an ontogenetical hypothesis by which phyletic memory is innate, while higher levels are the results of further cortical associations. Executive memory is so stored in the PFC: its lowest level will be the primary motor cortex, while the highest levels should represent more complex schemas and plans of goal-oriented actions. Because the execution of these schemas or plans requires the mediation of cross-temporal contingencies, PFC is supposed to be crucial in the temporal organization of behavior.

Complex behaviors require integration of both perceptive and executive hierarchies. To do it, long corticococortical fibers connect areas involved in these hierarchies. At each stage, upper frontal areas process global aspects of the sequence, while sensory signals occur. In this process sensory inputs from posterior cortex are progressively more concrete and more dependent on immediate temporal and spatial context.[16] There are some signals (episodic) that are processed in a wider temporal context that implies actions dependent from a high degree of temporal integration. In this case a simultaneus activation can be seen in the posterior cortex as in the rostral PFC. So, signals are processed at the same time in both cortices, being integrated with previous information (rules, instructional cue...) before lowering down. This *integration provided by PFC is not just across the time but is also cross-modal.*

A study on cross-modality in PFC have focused on the associations between visual stimuli and motor actions.[17] This study shows the evidence of the role of PFC in integrating visual and auditive stimuli across the time. *The paired association by PFC cells take place across modalities, across time and towards a goal.* Thus, Dorsolateral prefrontal and premotor cortices are involved on the menagement of temporal behavior, as motor sequences are. This role may be crucial also for the organization of the language that can be considered as a subset of motor sequence and that depends on the temporal integration of stimuli encoded in the two sense modalities.

3. Models of PFC functioning

Several neurophysiological studies on non-human primates and neuropsychological and neuroimaging researches on the task conditions under which PFC is engaged. However it is still missing an exhaustive understanding of the mechanisms of PFC control. For this reason bio-inspired models simulating the PFC can help us to understand better how the top-down cognitive control works through it. In this section will be given an account of main computational models contributes proposed in last years, focussing on models developed in order to understand the role of PFC in action sequences learning and performing.

3.1. *Main characteristics of PFC models*

Many computational studies had confirmed some hypotheses done[1] about the identification of neural correlates of plasticity in PFC, suggesting that these may operate as mechanisms for self-organization.[18] Some models of PFC functioning reproduce tasks of experiments in which subjects have to use game-rules, internal representations of goals, and means to achieve them.[7,8] Here, PFC is simulated as a hidden layer providing a bias getting stronger when there's a competition between automatic, strong stimulus-response mappings and controlled, weak one and favouring the seconds (e.g. model simulating the Stroop Task in[19]). This model even shows how an uninterrupted activity is necessary to improve a control mechanism. It still remains to show what happens in presence of a new task that requires a rapid updating of our PFC representations (e.g.[20]). The updating has to be, at the same time, adaptive and robust. In[18] has been proposed that Dopamine neurons (DA) may play a rule in this process gating the access to PFC by modulating the influence of its afferent connections. This process seems to be formally equivalent to what happens in many models that simulate the PFC updating mechanisms. Computational studies as the one of[18] confirm the plausibility of this self-organizing bootstrapping mechanism.

3.2. *PFC models and action sequences*

Within PFC studies inquiries on human motor learning behavior and action sequences have a particular relevance. Here we consider four computational model of particular interest, implemented to understand the role of PFC in controlling action sequences.

The first model reviewed is that by Gupta and Noelle.[7] The authors move from the hypothesis that there are two largely distinct neural pathways that control respectively the controlled and the automatic processes. The neural network is a model of the dual pathway hypothesis that uses the Leabra modeling framework[21] which incorporates two ways to modify the strength of connections: a) an error correction learning algorithm b) a Hebbian correlational learning rule. For the authors, the use of Leabra is strongly compatible with modeling a dual pathway model. The task reproduces some human experiments in which the subject had to learn sequences of key pressing on a keyboard of 9 keys. The network manages a two joint planar arm that has to press keys in sequence. The model includes a cognitive control modulation mechanism. This mechanism modulates the strength of the controlled pathways contribution to the final motor output as well as the strength of the input going from the controlled pathway to the automatic one. It is interesting to note that controlled pathway learns more rapidly (in terms of trials) than the automatic one. Automatic pathway in isolation cannot produce correct motor sequences. Beside, the controlled pathway is able to compensate the automatic pathway errors. At the last stage of learning, the model suffers when excessive control is employed during the execution of an automatized motor skill. The main limit of this model, as admitted by the authors themselves, is that it does not yet capture execution-time differences between controlled processing and automatic processing. It is well established that controlled execution of a skill is slower than automatic execution.

With the aim of understanding the top-down control exerted by internally generated sub-goal and by externally provided goal, Polk and colleagues[8] have developed a model to simulate the Tower of London TOL task. The simulation leads to a specific hypothesis about the role of the dorsolateral prefrontal cortex (DLPFC) in TOL: the DLPFC represent internally-generated subgoals that bias competition among choices toward the solution of the task. The TOL task involves moving three colored balls until they match a given goal configuration. An externally provided goal leads the system to prefer the goal-achieving move over the other legal moves. The model itself highlights the result of the combination of bottom up mechanism (or purely data driven production system) and a top down mechanism (a goal modulated system) in modern production systems. Though we still need a deeper comprehension of temporal organization mechanism and of PFC cooperation with other cortical areas.

The recent model by Botvinick and colleagues[4] is based on a computational framework, namely the actor-critic framework.[22] Such framework includes a hierarchical reinforcement learning (HRL)[23] to aggregate actions in subroutines that can be used as building block to solve incoming problems. Moreover the framework is endowed with temporally abstract actions, representations that cluster a set of interrelated action as a single higher level action or skills. These temporally abstract actions rather than specifying a single primitive action specify a whole policy to be accomplished, that is a mapping from states to actions. It is important to highlight that prefrontal representations do not implement policies directly but instead select among stimulus-response pathways implemented outside the PFC: in short, PFC working concerns the hidden layer. As the model shows a twofold relevance, neural and behavioural, an attempt has been made to map HRL on to functional neuroanatomy: a correspondence can be found between the actor and the dorsolateral striatum (DS) and between the critic with the ventral striatum (VS) and the mesolimbic dopaminergic system. Hence, representations within PFC correspond to option identifiers in HRL (an option being a sort of supergoal , e.g. prepare coffee, that calls lower-level options, e.g. adding sugar or cream), while stimulus response pathway selected correspond to option-specific policies. This mechanism can give an account of the role of the PFC to represent action as multiple, nested levels of temporal structure. Moreover it may find evidence in recent observation of primate behaviour: when cognitive planning involves a complex number of action sequences, cells in lateral PFC selectively exhibit an for a specific category of behavioral sequences. Categories of behaviors are embodied as sequences of movements and, during the planning, their representations are present in prefrontal cells. Authors identified not only cells dedicated to plan sequences, but also cells selective for the category of the sequences themselves. This implies the existence of a unit of knowledge that specifies the macrostructure of an event series at an advanced level of unification. It seems to be confirmed the theory of hierarchical structures of behavioral plan. This research confirm other evidences of the PFC role of categorization in monkeys.[6]

Finally the model described by Hazy and colleagues[3] adopts a radically different approach to understand how the PFC is involved in action sequences. The authors explicitly moves from the existing mechanistic models of the basal ganglia (BG) and frontal system. BG, in fact, provides a modulation of frontal action selection in terms of Go vs Not Go and this makes them play crucial role in motor control and action selection. Basal ganglia are responsible for learning by trial-and-error to automatically compose various sensorimotor primitives of the direct pathways, on the basis of a double inhibition mechanism in order to produce sophisticated behaviours. They are supposed to learn to select and compose sensorimotor skills on the basis of trial-and-error mechanisms, that can be mimicked by reinforcement learning algorithms.[22] However, once trained basal ganglia produce quite inflexible and stereotyped behaviour, elicited by just the right stimuli. Moreover, they do not generalize well to novel situations. However the two brain districts strongly interplay to produce voluntary behaviours. In this model the BG modulates working memory representations in prefrontal area. This allows to build on more abstract executive functions, as plans, goals, task-relevant stimuli, etc. The same mechanisms that allows the BG-PFC system to learn when to update or maintain its working memory informations can be extended to the output-gating mechanism. For these reasons authors implemented a PBWM (PFC, BG, Working Memory) model that is strongly bio-inspired and has the aim to give an account the strict relationship between the BG and PFC. The hypothesis of the authors is that PFC is an evolution of the BG and the frontal cortical system mechanisms. BG modulates PFC representations in terms of Go vs Not Go and this allows PFC to develop more abstract representations that are the ones stably maintained. The task of PBWM is to resolve the 1-2-AX task, an evolution of the simple AX task, that needs the model to answer to six key functional demands for working memory: rapid updating, robust maintenance, multiple separate working memory representations, selective updating, independent output-gating for top-down biasing of processing; Learning what and when to gate. The PVLV (Primary Value and Learned Value) moves away from the classical time differences (TD) learning mechanisms based on the predictive nature of dopamine activity (DA) and involves two learning mechanisms, separated but interdependent, based essentially on delta-rule. Another important aim of[3] is to elaborate a Multi-Task (MT) model, able to resolve several task and not just one as most of models early implemented. This means to reproduce our attitude to be flexible thanks of our ability in generalize, namely to abstract from specific situations. This model is just an extension of the PBWM. PFC an BG are implemented as layers of the same area interacting with the layers deputes to learn (Primary Value and Learned Value) through a reinforcement learning mechanism that simulates the midbrain dopaminergic system and its activation via the BG and the amygdala. The alghorithms used in the model are Leabra and kWTA (k-winners-take-all).

4. Discussion

PFC can be considered the central executive of our controlled behavior. We reviewed some of the most important theories that show how PFC is responsible for the flexibility of human behavior. If early theories and models[1] helped us to understand which were the general functions of PFC, some specific aspects were still to be cleared up. It remains to understand *how to capture execution-time differences between controlled processing and automatic processing*,[7] because it is clear that controlled execution of a skill is slower than an automatic one. We compared bio-inspired model[3] with models starting from the computational-framework model as the one proposed by.[4] We also noticed how the computational-framework[4] model may reproduce evidence in recent observation of primate behaviour: when cognitive planning involves a complex number of action sequences, cells in lateral PFC selectively exhibit an for a specific category of behavioral sequences.[6] There are still many opened problems as: how these models can quickly be updated when their encounter new cognitive tasks; how PFC is functionally organized; how does it work human capacity for generativity; which is the role of dopamine effects in PFC; how to give an account of *PFC interaction with specific subcortical areas*. We suggest to draw a biologically plausible model of PFC in order to implement the development of PFC in humans from childhood to adulthood. How does the shift work from the automatic to the controlled pathway in children that haven't still developed PFC? We are going to keep on searching about that.

References

1. E. K. Miller and J. D. Cohen, *Annual Review of Neuroscience* **24**, 167 (2001).
2. J. M. Fuster, *Neuron* **30**, 319 (2001).
3. T. E. Hazy, M. J. Frank and R. C. O'Reilly, *Philosophical Transactions of the Royal Society B* **362**, 1601 (2007).
4. M. Botvinick, *Trends in Cognitive Sciences* **12**, 201 (2008).
5. E. K. Miller, D. J. Freedman and J. D. Wallis, *Philosophical Transactions of the Royal Society B* **357**, 1123 (2002).
6. K. Shima, M. Isoda, H. Mushiake and J. Tanji, *Nature* **445**, 315 (2007).
7. A. Gupta and D. Noelle, A dual-pathway neural network model of control relinquishment in motor skill learning, in *Proceedings of the International Joint Conference on Artificial Intelligence*, 2007.
8. T. Polk, P. Simen, R. Lewis and E. A. Freedman, *Brain Research Cognitive Brain Research* **15**, 71 (2002).
9. M. A. Goodale and A. D. Milner, *Trends in Neuroscience* **15**, 20 (1992).
10. J. P. Ray and j. L. Price, *Journal of Comparative Neurology* **337**, 1 (1993).
11. J. O'Doherty, E. T. Rolls, R. Bowtell and e. a. McGlone, F., *NeuroReport* **11**, 893 (2000).
12. E. Koechlin, *The cognitive architecture of the human lateral prefrontal cortex*, in *Attention and Performance*, ed. R. Y. K. M. Haggard, P. (Psychological Bullettin, 2008), pp. 273–293.
13. T. Paus, *Trends in Cognitive Sciences* **9**, 60 (2005).
14. M. Knopf, R. A. Kressley-Mba, L. T. f. d. i. o. . Simone and . step action sequences with 6-month-olds. Infant Behavior & Development 28 8286 (2005)., *Infant Behavior and Development* **28**, p. 8286 (2005).
15. J. A. Sommerville and A. L. Woodwardb, *Cognition* **95**, 1 (2005).
16. J. Fuster, *Trends in Cognitive Science* **8**, 143(Apr 2004).
17. J. Fuster, M. Bodner and J. Kroger, *Nature* **405**, 347(May 2000).
18. T. Braver and J. D. Cohen, A computational model of prefrontal cortex function, in *Advances in Neural Information Processing Systems*, ed. T. L. DS Touretzky, G Tesauro, *A* (MIT Pres, 1995).
19. J. D. Conhen and J. L. McClelland, *Psychological Review* **97**, 332 (1990).
20. S. Dehaene and J. P. Changeux, *Cerebral Cortex* **1**, 62 (1989).
21. R. C. O'Reilly, *Neural Computation* **8**, 895 (1996).
22. A. G. Barto, Adaptive critics and basal ganglia, in *Models of Information Processing in the Basal Ganglia*, ed. D. B. JC Houk, J Davis (MIT Press, 1995).
23. R. S. Sutton, D. Precup and S. Singh, *Artificial Intelligence* **112**, 181 (1999).

A NEURAL-NETWORK MODEL OF THE DYNAMICS OF HUNGER, LEARNING, AND ACTION VIGOR IN MICE

ALBERTO VENDITTI, MARCO MIROLLI, DOMENICO PARISI,
GIANLUCA BALDASSARRE

Istituto di Scienze e Tecnologie della Cognizione,
Consiglio Nazionale delle Ricerche (ISTC-CNR)
Via San Martino della Battaglia 44, I-00185 Roma, Italy
{alberto.venditti, marco.mirolli, domenico.parisi, gianluca.baldassarre}@istc.cnr.it

Recently the computational-neuroscience literature on animals' learning has proposed some models for studying organisms' decisions related to the energy to invest in the execution of actions ("vigor"). These models are based on average reinforcement learning algorithms which make it possible to reproduce organisms' behaviours and at the same time to link them to specific brain mechanisms such as phasic and tonic dopamine-based neuromodulation. This paper extends these models by explicitly introducing the dynamics of hunger, driven by energy consumption and food ingestion, and the effects of hunger on perceived reward and, consequently, vigor. The extended model is validated by addressing some experiments carried out with real mice in which reinforcement schedules delivering lower amounts of food can lead to a higher vigor compared to schedules delivering larger amounts of food due to the higher perceived reward caused by higher levels of hunger.

Keywords: Fixed and random ratio schedules, neural networks, average reinforcement learning, motivations, needs, energy costs, phasic and tonic dopamine

1. Introduction

The action of dopamine neuromodulation is believed to exert a powerful influence on *vigor*, that is the strength or rate of responding in behavioural experiments. There are many psychological theories that attribute the vigor effects to a variety of underlying psychological mechanisms, including incentive salience,[1,2] Pavlovian-instrumental interactions,[3,4] and effort-benefit tradeoffs.[5] A different line of research, using the electrophysiological recording of midbrain dopamine neurons's activity in awake behaving monkeys, suggests that the phasic spiking activity of dopamine cells reports to the striatum a specific "prediction error" signal.[6–9] Computational models have shown that this signal can be used efficiently both for learning to predict rewards and for learning to choose actions so as to maximize reward intake.[10–14] However, these theories have some important limitations. First, they only try to explain choice between discrete actions whilst they do not say anything about the strength or vigor of responding. Second, they generally assume that dopamine influences behaviour only indirectly by controlling learning whereas dopamine might have other effects on behaviour. Finally, they are only concerned with the phasic release of dopamine, while the tonic level of dopamine constitutes a potentially distinct channel of neuromodulation that might play a key role in energizing behaviour.[15,16]

Niv et al.[17] proposed a normative account of response vigor which extends conventional reinforcement learning models of action choice to the choice of vigor, that is to the energy expenditure that organisms associate to the execution of chosen actions. To pursue this goal the authors use a model of learning different from the model normally used to study phasic dopamine and reward prediction error, namely the *actor-critic* model based on the *Temporal Difference learning rule*.[18] Rather, they use an actor-critic model based on the *average rate of reward*. The average rate of reward exerts significant influence over overall response propensities by acting as an *opportunity cost* which quantifies the cost of sloth: if the average rate of reward is high, every second in which a reward is not delivered is costly, and therefore actions should be performed faster even if the energy costs of doing so are greater. The converse is true if the average rate of reward is low. In this way the authors show that optimal decision making on vigor leads to choices with the characteristics of choices exhibited by mice and rats in behavioural experiments.

Notwithstanding its pioneering value, the work of Niv et al.[17] has two limits which are addressed here. First, it does not study how food's reinforcing value is influenced by the dynamics of internal needs, e.g. *hunger*. Second, it studies only the steady state values of variables and not their dynamics during learning. This paper proposes a computational model which includes a sophisticated internal regulation of hunger and allows investigating behaviour *during* learning.The results are compared with data from experiments carried out with real mice by Parisi.[19]

The rest of the paper is organised as follows. Section 2 descrives the targeted experiments. Section 3 illustrates the model and the simulated mice. Section 4 compares the behaviour of simulated and real mice. Finally, Section 5 draws the conclusions.

2. Target experiments

Parisi[19] tested 36 mice in a linear corridor at the end of which they could find a pellet of food, and measured the time they employed to reach the end of the corridor. Here we interpret the speed of mice as an indicator of the vigor invested in the execution of actions. Food was delivered according to three different schedules of reinforcement to three different groups of mice: (a) Fixed Ratio 100% (FR100): food was always delivered when the corridor end was reached. (b) Fixed Ratio 50% (FR50): food was delivered only in odd trials. (c) Random Ratio 50% (RR50): food was delivered randomly with a probability of 50%.

Figure 1a shows the mice's speed curves during learning along various days of training (for each day the average performance for 6 trials is reported). After each daily session the mice had free access to food for half an hour and then were kept without food until the next day session. Figure 1b shows in detail the speed of mice related to FR50 and RR50, separately for trials with and without food (respectively denoted with FR50+ and FR50-. The graphs show that: (a) Mice trained with RR50 exhibited the highest level of vigor, followed by the mice trained with FR100 and then by those trained with FR50 (lowest vigor). The first goal of this paper is to explain why FR100 led to a higher level of vigor with respect to FR50. The high level of vigor exhibited by mice with RR50 was probably caused by some energizing effects of the randomness of action outcomes and will not be further discussed in the paper. (b) Figure 1b shows that FR50+ led to a vigor lower than FR50-. Parisi explained this result suggesting that the reward not only affects learning but it also allows mice to predict the outcome of the succeeding trial (notice that this can happen in FR50, as trials with and without reward alternate and so are predictable, but not in RR50). A second goal of the paper is to validate this hypothesis with the model. (c) Figure 1b also show that FR50+ led to a vigor higher than FR100. At first sight, this is counterintuitive as the reward in FR50+ and FR100 trials is identical. A third goal of the paper, the most important one, is to explain this result in terms of dynamics of *hunger*, namely the fact that higher levels of hunger can increase the perceived reward associated with food. (d) Figure 1b also shows that before vigor levels reach a steady state, FR50- produces the highest levels of vigor, in particular higher than FR50+ and FR100. Parisi explained this by saying that the trials related to FR50- were those taking place right after a rewarded trial (FR+ series). The fourth goal of the paper is to specify and integrate this explanation.

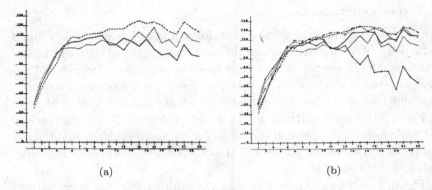

(a)   (b)

Fig. 1. Results of the target experiments. In both graphs, the x-axis refers to successive groups of 6 trials and the y-axis refers to mice's speed (vigor) measured as 100 divided by the number of seconds spent to cover the corridor. (a) The evolution of speed during learning with the three schedules of reinforcement: the highest dashed curve refers to RR50, the intermediate dotted curve to FR100, and the lowest continuous curve to FR50. (b) Same data with separated curves for FR50+ and FR50- (highest dashed curves), and for RR50+ and RR50- (continuous curves); the curve of FR100 is the same.

Indeed, a further explanation is needed beyond that of Parisi as *both* FR50- and FR100 conditions involve trials following rewarded trials.

3. The model

3.1. *The task*

The simulated environment is composed by a corridor measuring 1.5 meters. In the experiments, the simulated mouse is placed at the left end of the corridor and is required to decide the speed (vigor) with which to move to the right end. When the mouse reaches the right end it can eventually find (and "eat") a reward (a unit of food) and then is replaced at the start position. The food is delivered according to one of the three reinforcement schedules illustrated in Section 4.

3.2. *The actor-critic component of the model*

The model is based on a neural-network implementation of the actor-critic reinforcement learning model[18] composed of two parts: the *actor* and the *critic* (in its turn mainly formed by the *evaluator*). In general the model is capable of learning to select appropriate actions in order to maximise the *sum of the future discounted rewards*: the evaluator learns to associate evaluations with visited states on the basis of the rewards experienced after these visits; the critic produces a *one-step judgment* of the actor's actions on the basis of the evaluations of couples of states visited in sequence; the actor learns to associate suitable actions with the perceived states of the environment on the basis of the critic's judgment.

This model has been chosen, among the several available reinforcement-learning models, because it has a considerable biological plausibility.[20] In particular, the model has several correspondences with the anatomy and physiology of the basal ganglia, which are deep nuclei of vertebrates' brain playing a fundamental role in action selection.[21]

The model is now illustrated in detail. The model has tree input units: (a) the first two units implement a memory of the outcome, in terms of reward, obtained in the previous trial (in particular, the units are activated with $< 1, 0 >$ when the rat has consumed food in the preceding trial, and with $< 0, 1 >$ otherwise); (b) the third unit is a *bias unit* always activated with 1.

The *actor* is a two-layer feed-forward neural network formed by the three input units, denoted with x_i, and by a sigmoidal output unit ranging in $[0, 1]$ and indirectly encoding vigor. The activation of the output unit is used as the centre μ of a Gaussian probability density function σ having standard deviation ς (set to 0.1) which is used to draw a random number that represents the chosen vigor:

$$\mu = \frac{1}{1 + \exp^{[-\Sigma_i w_{ai} \cdot x_i]}} \qquad y \sim \sigma[\mu, \varsigma] \qquad (1)$$

where w_{ai} are the actor's weights from the input units x_i to output unit y and "$\sim$" indicates the probability density function of y (the Gaussian's tails are cut at 0 and 1 by redrawing new numbers when this range is violated). The action y (the selected vigor) is drawn randomly "around μ" as reinforcement learning models need a certain randomness to find suitable solutions by trial-and-error. The activation of the output unit of the actor is used to set the mouse's speed (a maximum vigor of 1 corresponds to a mouse's step size measuring $1/10$ of the corridor length).

The *evaluator*, which is part of the *critic*, is a network that uses the activation of the three input units of the model to return, with its linear output unit, an estimation of the theoretical evaluation of the world state corresponding to the input pattern. The "theoretical evaluation" to be estimated, V, is defined as the sum of the future discounted rewards each decreased of the average per-step long-term reinforcement:[22-25]

$$V[t] = E_\pi \left[\sum_{k>t} [R[k] - \overline{R}] \right] \qquad (2)$$

where E_π is the expected sum of future rewards averaged over the possible actions selected by the current action policy π expressed by the current actor, R is the reinforcement, and $\overline{R}$ is the average (per-step) long-term reinforcement. Note that, as suggested by Niv et. al,[17] $\overline{R}$ might be thought to correspond to the tonic dopamine level, encoding the opportunity cost of each time unit engaged in any activity. With this respect, it is important to notice that many experiments show that high levels of striatal dopamine are strongly associated with an high rate of response, that is vigor.[15,16] Interestingly, this happens even before *phasic dopamine* underlying learning (and corresponding to the model's surprise $S[t]$ illustrated below) has a full effect on action selection.[2] In the simulations, $\overline{R}$ is estimated on the basis of the experienced past rewards R. The evaluator produces an estimation $\hat{V}$ of the theoretical V:

$$\overline{R}[t] = (1 - \kappa)\overline{R}[t - 1] + \kappa R[t] \qquad \hat{V}[t] = \sum_i [w_{vi}[t]x_i[t]] \qquad (3)$$

where w_{vi} are the evaluator's weights ($0 < \kappa < 1$ was set to 0.01).

The *critic* computes the surprise $S[t]$ used to train (as illustrated below) the evaluator to produce increasingly accurate $\hat{V}$ and the actor to produce actions leading to increasingly high and/or frequent rewards:

$$S[t] = \left(R[t] - \overline{R}[t] \right) + \hat{V}[t] - \hat{V}[t - 1] \qquad (4)$$

The evaluator uses the Temporal Difference algorithm (TD[18]) to learn accurate estimations $\hat{V}$ with experience as follows:

$$w_{vi}[t] = w_{vi}[t - 1] + \nu \cdot S[t] \cdot x_i[t - 1] \qquad (5)$$

where ν is a learning rate (set to 0.2).

The surprise signal is also used by the actor to improve its action policy. In particular, when surprise is positive, the centres of the Gaussian functions used to randomly draw the vigor level are made closer to the actually drawn value, whereas when surprise is negative such centre is moved away" from it. This is done by updating the actor's weights as follows:

$$w_{ai}[t] = w_{ai}[t-1] + \zeta \cdot S[t] \cdot (y[t-1] - \mu[t-1]) \cdot (\mu[t-1](1 - \mu[t-1])) \cdot x_i[t-1] \quad (6)$$

where $(\mu[t-1](1-\mu[t-1]))$ is the derivative, with respect to the activation potential, of the actor sigmoid output units' activation, ζ is a learning rate (set to 0.2), $(y[t-1] - \mu[t-1])$ is the part of the formula that moves the centres of the Gaussian towards, or away from, the noisy vigor selected by the actor when surprise $S[t]$ is respectively positive or negative. The motivation behind this updating rule is that a positive surprise indicates that the action randomly selected by the actor at time $t-1$ produced reward effects at time t better than those expected by the evaluator at time $t-1$: this means that such drawn action is better than the "average action" selected by the actor at time $t-1$, as estimated by the evaluator, and so such action should have an increased probability of being selected in the future in correspondence to $x_i[t-1]$. A similar opposite reasoning holds when surprise is negative.

3.3. *The dynamics of costs, hunger, and perceived rewards*

This section illustrates the novel part of the model related to the simulated mouse's energy need (hunger), the energy costs caused by action vigor, the resulting energy balance, and the effects of this on the reward that the mouse perceives when it eats the food. The model of Niv et al.[17] already considered a structure of costs similar to the one illustrated below; however, it did not consider hunger and its effects on perceived rewards, as done here.

In every step, the mouse incurs in two types of costs: (a) a fixed unitary (i.e. per step) cost FUC, set to 0.01; (b) a variable unitary cost VUC set to a maximum level of 0.99: this cost is modulated by the vigor y to capture the fact that more vigor spent executing actions implies a higher energy cost. The sum of the two costs gives the total unitary costs TUC. The energy level E varies on the basis of the energy costs and food ingestion:

$$TUC = FUC + VUC \cdot y^\iota \quad E[t] = E[t-1] + \varepsilon \cdot F[t] - \chi \cdot TUC \quad (7)$$

where ι is a exponential parameter (set to 5.0) implying that costs grow more than proportionally when vigor grows; ε is the energy increases due to the ingestion of one unit of food (set to 0.01), F indicates the units of food ingested when the reward is delivered (set to 10), χ is the decrease of energy due to energy costs (set to 0.05). E is always kept in the range [0, 1]. Moreover, and importantly, at the end of each block of six trials (corresponding to a day session) E is set to 0.2 to represent the fact that after each trial the real mice had free access to food and then were kept without food until the succeeding day session.

Hunger H depends on the level of energy. The perceived reward R, which drives the learning processes of the actor-critic model's components, depends not only on the ingested food but also on the hunger level that modulates the appetitive value of food:

$$H[t] = (1.0 - E[t])^{\varphi} \quad R[t] = F[t] \cdot H[t] \tag{8}$$

where φ is a parameter (set to 3.7) that causes an exponential increase of hunger in correspondence of lower levels of energy.

Figure 2 shows the mouse's costs, perceived rewards, and their balance (difference), all measured per time unit, in correspondence to different levels of vigor and assuming that the mouse starts to run along the corridor with a maximum energy level. The unitary perceived reward UR used to plot the curves was obtained as follows:

$$UR = (F \cdot H) / (1.5 / (MS \cdot y)) \tag{9}$$

where MS is the maximum step size of the mice (set to 1/10 of the corridor length, that is to 0.15), corresponding to the maximum vigor ($y = 1$), and $(1.5 / (MS \cdot y))$ represents the number of steps needed by the mice to cover the corridor length (1.5 m) with a vigor y.

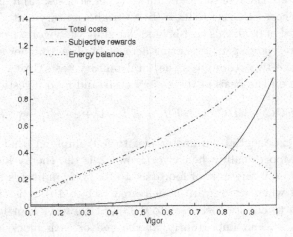

Fig. 2. The curves represent the energy costs, the perceived rewards, and the energy balance in relation to increasing levels of vigor (x-axis).

Consider that due to the small duration of a trial the energy spent in terms of TUC is rather low whereas the energy expenditure related to the time that elapses from one day to the following one, which brings E to 0.2 (see above), is rather high and causes the most important effects on perceived rewards. In any case, the dynamics of costs (FUC, VUC and TUC) were included to propose a general model with the potential of tackling many experiments involving hunger. In this respect, it is important to mention that graphs as the one reported in Figure 2, and an analysis of costs as the one reported in this Section, resemble those used by economists to analyse the costs, income and balance of enterprises. Indeed, a recent interdisciplinary research trend in neuroscience aims to exploit the analytical tools used by economics to investigate phenomena related to the functioning of brain.[26] The analysis reported in this section, for example, allowed us to conduct a preliminary exploration of some of the parameters of the model so as to be able to identify interesting regions of them (e.g. this allowed us to envisage the possible vigor value which could maximise the energy balance: see the maximum value of the energy-balance curve in Figure 2).

4. Results

The model was implemented in Java programming language and was tested five times for each of the three reinforcement schedules FR100, FR50 and RR50. The curves of Figure 3 show the level of vigor selected by the model in the three conditions during 10,000 trials of learning. The vigor for FR50 is also plotted for rewarded (FR50+) and non-rewarded (FR50-) trials. These results are now compared with those of Figures 1a-b concerning real mice.

The first result of the simulation is that, as in real mice, with FR100 the simulated mouse selects a level of vigor higher than with FR50. This is likely due to the higher overall energizing effect due to the higher amount of food ingested. More importantly, the model succeeds in reproducing the behaviour of real mice that exhibit a higher vigor with FR50+ than with FR50-: as suggested by Parisi, the reward not only effects learning but, being stored in the model's memory input units, it can also play the role of predictor of the outcome of the next trial.

Figure 1b shows that in real mice FR50+ led to a vigor higher than FR100. As mentioned in Section 2, this result is unexpected as in the two conditions the reward is the same, namely 10 units of food. The model, which reproduces this outcome, allows explaining the mechanism behind it. In the model each group of six trials (corresponding to a "day section" of the experiments with real mice) starts with a level of energy of 0.2. Even

if in FR50+ in three trials out of the six of each block the level of energy
increases, on average when food is ingested the level of hunger is higher
than in FR100. As high levels of hunger increase the *perceived* reward, the
mouse learns to spend more energy to get one unit of food in FR50+ than
in FR100. Notice how this mechanism might have an adaptive value in
ecological conditions as it leads mice to spend more energy when food is
scarcer and the risk of death for starvation is high.

Interestingly, the model also reproduces the behaviour exhibited by real
mice for which in early phases of learning FR50- produces levels of vigor
higher than in the other conditions, in particular FR100 and FR50+. Parisi
explained this noticing that the trials related to FR50- were those taking
place right after a rewarded trial from FR+. The model suggests detailed
mechanisms behind this explanation. According to what stated in the previous paragraph, FR50- trials follow the receiving of the highest perceived
reward. In FR50, before the mouse learns to predict if a trial will be rewarded or not the connection weight related to the bias unit will tend to
increase maximally in rewarded FR50+ trials and so to contribute to a
high vigor in the following FR50- trial. In FR100 this effect is lower as the
perceived reward is lower.

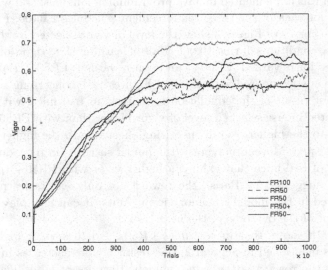

Fig. 3. Levels of vigor during learning, lasting 10,000 steps, in the conditions FR100,
RR50, FR50, FR50+ and Fr50-. Each curve is an average of five repetitions of the
simulations.

5. Conclusion

This paper presented a preliminary study of a model that extends the work of Niv et al.[17] concerning the level of vigor with which animals execute actions by introducing explicitly the dynamics of *hunger* and its effects on *perceived rewards*. The extension makes it possible to reproduce most of the results obtained by Parisi[19] with real mice experiments. The model explains various aspects of the behaviours exhibited by real mice in terms of specific mechanisms, in particular the fact that the vigor of an action can be high even in the presence of low amounts of food received if high levels of hunger lead the mice to perceive food as more rewarding.

The model, however, was unable to reproduce the result according to which real mice trained with RR50 are faster than those trained with FR50 and FR100. As mentioned in Section 2, this particular behaviour is likely due to other mechanisms not taken into consideration by the model, in particular the possible energizing effects of *random* uncertain outcomes. These energizing effects might have adaptive value as they would lead animals to further explore the environment to collect more information and decrease uncertainty. This topic should be addressed in future research.

A second, more important limit of the model, shared with the model of Niv et al.,[17] is that it performs a choice of vigor which is "cognitive", that is, it is learned and implemented on the basis of reinforcement learning mechanisms underlying the selection of actions themselves. On the contrary, probably the nervous system of animals contains some mechanism specifically dedicated to controlling the level of energy invested in actions' performance. One result suggesting that this might be the case is the fact that the model learns to regulate the level of vigor very slowly (in about 4,000 trials) while real mice regulate the level of vigor after few trials, often even before they learn to produce the correct action.[17] Also this issue should be tackled in future work.

Acknowledgments

This research was supported by the EU Project *ICEA - Integrating Cognition, Emotion and Autonomy*, contract no. FP6-IST-IP-027819. Venditti, Mirolli, and Baldassarre consider this research also a small homage to the bright and curious mind of the young Domenico Parisi which, 43 years ago, in times of behaviourism dominance, led him to go abroad in USA to carry out the experiments addressed in the paper. This experience was one of the first forward-running steps of a long, always-innovative, enthusiastic scientific career which ultimately arrived to mark forever the life of the co-authors and other dozens of young scientists.

References

1. K. C. Berridge and T. E. Robinson, *Brain Res Brain Res Rev* **28**, 309 (1998).
2. S. Ikemoto and J. Panksepp, *Brain Res Brain Res Rev* **31**, 6 (1999).
3. A. Dickinson, J. Smith and J. Mirenowicz, *Behav Neurosci* **114**, 468 (2000).
4. A. Murschall and W. Hauber, *Learn Mem* **13**, 123 (2006).
5. J. D. Salamone and M. Correa, *Behav Brain Res* **137**, 3 (2002).
6. T. Ljungberg, P. Apicella and W. Schultz, *J Neurophysiol* **67**, 145 (1992).
7. W. Schultz, P. Apicella and T. Ljungberg, *J Neurosci* **13**, 900 (1993).
8. W. Schultz, *J Neurophysiol* **80**, 1 (1998).
9. P. Waelti, A. Dickinson and W. Schultz, *Nature* **412**, 43 (2001).
10. R. S. Sutton and A. G. Barto, *Psychol Rev* **88**, 135 (1981).
11. K. J. Friston, G. Tononi, G. N. Reeke, O. Sporns and G. M. Edelman, *Neuroscience* **59**, 229 (1994).
12. A. G. Barto, *Curr Opin Neurobiol* **4**, 888 (1994).
13. P. R. Montague, P. Dayan and T. J. Sejnowski, *J Neurosci* **16**, 1936 (1996).
14. W. Schultz, P. Dayan and P. R. Montague, *Science* **275**, 1593 (1997).
15. G. D. Carr and N. M. White, *Pharmacol Biochem Behav* **27**, 113 (1987).
16. D. M. Jackson, N. E. Anden and A. Dahlstroem, *J Psychopharmacol* **45**, 139 (1975).
17. Y. Niv, N. D. Daw, D. Joel and P. Dayan, *J Psychopharmacol* **191**, 507 (2007).
18. R. Sutton and A. Barto, *Reinforcement Learning: An Introduction.* (MIT Press, Cambrige, MA, USA, 1998).
19. D. Parisi, Il rinforzo come stimiolo discriminativo, in *Atti del XV Congresso degli Psicologi Italiani*, 1965.
20. J. Houk, J. Davis and D. Beiser, *Models of Information Processing in the Basal Ganglia* (MIT Press, Cambridge, MA, USA, 1995).
21. P. Redgrave, T. J. Prescott and K. Gurney, *Neuroscience* **89**, 1009 (1999).
22. A. Schwartz, A reinforcement learning method for maximizing undiscounted rewards, in *Proceeding of the Tenth Annual Conference on Machine Learning*, 1993.
23. S. Mahadevan, *Machine Learning* **22**, 159 (1996).
24. J. Tsitsiklis and B. V. Roy, *Automatica* **35**, 1799 (1999).
25. N. D. Daw and D. S. Touretzky, *Neural Comput* **14**, 2567 (2002).
26. P. Phillips, M. Walton and T. Jhou, *J Psychopharmacol* **191**, 483 (2007).

BIO-INSPIRED ICT FOR EVOLUTIONARY EMOTIONAL INTELLIGENCE

MARCO VILLAMIRA

Institute of Human, Language and Environmental Sciences, IULM University – Milan
Via Carlo Bo, 8, 20143, Milan, Italy

PIETRO CIPRESSO

Institute of Human, Language and Environmental Sciences, IULM University – Milan
Via Carlo Bo, 8, 20143, Milan, Italy

In this work, we deal with the possible application of an agent-based system for evolutionary emotional intelligence. Starting with hypotheses on the evolutionary history of emotions (even mythological hypotheses concerning cosmogony), we try to define emotions operatively, considering the great problem of highlighting emotions vital for the survival of individuals and species. Here, the psychological and semantic problem is establishing an effective operative meaning for 'emotions'. The concept of emotional intelligence then is introduced. This concept is partially contradictory, invoking two terms (emotion and intelligence) that – in our culture – have had virtually opposite connotations for centuries. Finally, we consider potential relationships between artificial agents and emotions.

1. Evolutionary history of emotions

1.1. Living systems

We could say that... in the beginning, there were structures of very low complexity; and than their main, and perhaps only problem, was to exist (see 'Transgression theory', M. A. Villamira, Comunicare, FrancoAngeli, Milano, in press). They then diffused and began to evolve. For a long time, the behaviors of 'Living Matter' were rather rudimentary.

144

In this context 'Transgression' is a behavior diverging from probability.

The meaning essentially is similar in the social sciences and also in jurisprudence, if we replace the term 'maximum probability' with 'normality'.

Transgression, also in this case, implies a "coming out from established rules and paths" in order to approach new areas, where there are even other rules in force.

Figure 1: ... from non-living matter ... complex molecules, protozoans ... the essential task of the cell membrane is generating a sort of 'individuality', multicellular beings and the generation of organs and apparatuses ... the specialization and the nervous system coming out ...

MYTHOLOGIC HYPOTHESIS IN COSMOGONY

→ *... at the beginning was Nothingness and in Nothingness Entropy governed ... but there is a rule, not yet explained, under which naught, never Nothingness, can last forever – if, outside our Universe, there is logic to speak about time - : so Entropy needed one's own contrariness, and Nothingness needed Something existing... tension grown ... and then blown up (...the Big Bang)...*

→ *Contradictions of our Universe were born including, as is its essence, All and Nothing, Probability and Improbability;*

→ *The biggest improbability, so the biggest among transgressions, was Life, placed in a low entropy area, producing and needing order, organization and variety;*

→ *Our Universe is contradictory and probably, in a contradictory way, needs contradiction and – despite this – its own basic rules exclude and punish it;*

→ *So it happens that some situations – such as life – exist, maybe necessarily (philosophically and thermodynamically speaking), but – contemporaneously – should not subsist, according to the basic rules of this Universe.*

→ *So it happens, as in social organizations, what should not happen, such as any kind of deviance, that however happens for a 'necessary' event, as above said, may last for a limited time (the time of our life or the time of our organizations) and then, obeying more general rules, go back to the rules; i.e. to die.*
Also it happens that living beings always have equipped themselves to improve and to extend one's own life, as well as to promote their existence and the diffusion of one's own species.

→ *So it is possible – and maybe compulsory – for rules and transgressions to exist and coexist, allowing this Universe to evolve dynamically ...*

Supposedly, each Living Being interacts with others, more or less different from it, 'simply' via avoidance, escape and/or pursuit mechanisms – such as the 'thinking vehicles' proposed by Valentino Braitenberg – and nevertheless already including, like Braitenberg's 'vehicles', *in nuce*, the opportunity for more complex evolutions ...

... going on tendentially slowly and inconstantly, transgressing from probability's rules, extended and yielded to complexity; new emergences lead to more and more complex situations and this generated other emergences ...

1.2. Emotions

An operative definition of 'emotions' :
We think that emotions could be defined as *'facilitating and/or induction tools of behavioral shortcuts'*.
This happens, above all, for those emotions that are linked with instant survival (fear → escape and anger → attack) through shortcuts connecting sense organs, cerebral areas and motor apparatuses.
About emotions not strictly linked with instant survival, the problem is considerably more complex, because of what we call 'emotions': many psychological states and situations marked with different meanings, sometimes distant from each other, both neuro-psychologically and operatively.
So it is very difficult to model and represent emotions through agents, situations, behaviors and, above all, their purpose and consequences, if it is not possible to have a clear vision of 'reality'.
One of the most critical tasks, therefore, is to identify relevant and operative definitions that lead us to an exact representation of what we call emotions. Certainly the growth of sciences (e.g., the mathematical sciences, from infinitesimal calculus, onward) is to represent with convenient approximation and deal with sufficient accuracy even in 'fuzzy' situations. Here, it is the ability of the researcher:

1) To define exactly the object of an experiment;
2) To choose the most suitable methodologies to deal with that object;
3) To be intellectually honest when communicating the results.

New chances impose new solutions, and basic mechanisms are not sufficient to manage situations that are becoming more and more complex; the time of evolution and the time of behavior selects the increased fitness endowed to individuals... and develops new mechanisms to make more effective escape and attack behaviors → fear and anger ... the basic structures of EMOTIONS.

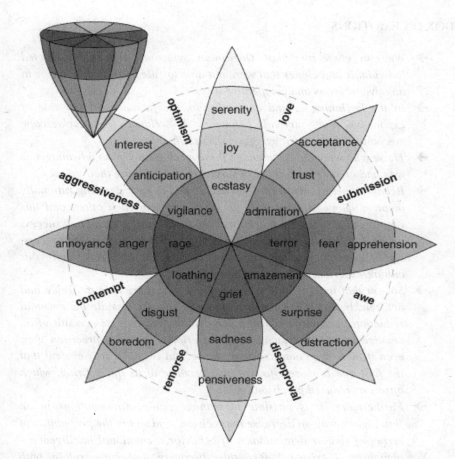

Figure 2: We think it interesting to consider a figure, the result of one of the many attempts to classify emotions and their composed effects (Robert Plutchik's psycho-evolutionary theory of emotions, 2002). It is obvious that difficulties arise when trying to establish exact bounds and clear effects concerning the different operative definitions.

BOX ON EMOTIONS

→ *We can conjecture that Darwinian selection tendentially deleted individuals and clones that were not able to rule their own behavior in an advantageous and competitive way.*

→ *In the beginning ... and currently too, the fundamental key side of expansion and survival behaviors is related to attack/escape mechanisms, sustained by aggressiveness/fear.*

→ *Those who were able to manage these mechanisms in an advantageous way had greater probability of surviving and diffusing their genes.*

→ *Behavior means, essentially, 'to decide and to act': deciding rationally involves many neurons in areas often far from each other, and this requires time, often too much time, compared with survival and success needs. Besides, it is not sure that acting rationally can lead to success, since our intellectual endowment is not perfect, and our culture/experience cannot be exhaustive.*

→ *So - at that time as well as now – those who were able to decide and act quickly had a greater probability of success, despite the eventual reduction in 'accuracy' when one decides and acts as a result of an emotive input (like machine gun superiority versus a precision gun; even if there are contexts where the second is better, it is evident that the first one is more effective, just because of its quick firing, which offsets its reduced precision).*

→ *Furthermore, it is possible to manage one's own emotions in an 'intelligent' way, to decrease imprecision' and so too the probability of errors of similar determinants of behavior... emotional intelligence – obviously – existed before this discovery and can explain both individual and social groups successes.*

→ *To give an example (humor inspired) we can refer to Anglo-Saxon culture that, at least since the Second World War, dominates our Planet. Many people think that the Anglo-Saxon approach to problems is the most cool and rational, but it is necessary to understand their society and their cultural roots (e.g., their humor) to understand that – de facto – this is EMOTIONAL INTELLIGENCE, i.e., the intelligent management of emotions ... so it is no accident that emotional intelligence is a product of Anglo-Saxon culture.*

1.3. Emotional intelligence

Emotional intelligence, definition:
'Emotional intelligence' remembers *'festinare lente'*, a kind of provocative oxymoron (a figure of speech that combines two normally contradictory terms).

At least since illuminism, our cultural tradition, has delineated almost impassable bounds between reason, which is the noble expression of intelligence, and emotions, which are the expression of something of visceral and uncontrolled instinct or passion.

On one hand, there is noble reason, the product of the superior mind of human beings; on the other hand are the emotions, a dangerous and ignoble heritage of animal world. This idea pervaded up to almost the end of the last century, throughout all of Western culture.

Towards the end of the last century, worldly-wise researchers began to doubt the foundations of this belief. One example, albeit apparently marginal but of fundamental importance to challenge old beliefs, has been the discovery of the low credibility and predictive power of standard intelligence tests (e.g., the IQ test), in terms of foreseeing professional and social success. How was it possible that subjects who were designated 'more intelligent than average' registered such minimal success in these areas? This problem relates directly to our issue, since the emphasis of this paper, in fact, is on subject performance, and the influences on the successes of subjects and their genetic clones.

We can try to define 'emotional intelligence' as the ability to manage 'intelligently' one's own emotions and the emotions of others too, that – *de facto* – supposes an optimal mix of emotionality and rationality, which certainly is difficult for Cartesian fundamentalists, engaged as always to divide the world into incompatible opposites; nonetheless, it surely would be appreciated by complexity lovers.

However, the fact remains that we must deal with concepts and situations that are only 'slightly exact'. Consequently, we would like to propose taking into consideration only those emotions that are directly connected to survival and success that, even if they have no specified starting points, nonetheless offer the opportunity to work in a less fuzzy context.

We wish to be able to improve our research skills as much as we can, progressively and ultimately to extend to all areas of emotion.

2. Simulating emotion

The primary aim of this paper is not to create a model for a right and operative simulation of such a complex psychobiologic process, but to generate food for thought in order to carry out, in terms of models, complex psychological structures that are not always manageable using informatics or mathematical instruments. Many scholars have analyzed decision-making processes, above all in economics. It is important to cite Daniel Kahneman "for having integrated insights from psychological research into economic science, especially concerning human judgment and decision-making under uncertainty" and Herbert Simon, who opposed to the model of 'Olympic rationality', the concept of *bounded rationality*.

Before speaking about emotions, in human or in artificial beings, we need to understand the context within these act. In temporal terms (but this is not the only axis to consider), we speak about antecedents and consequents:

- *antecedents* are the environment and its occurrences, as well as emotional elicitation and any associated attentive processes (strongly influenced by the emotional sphere);
- *consequents* are behavioral shortcut processes that exert a direct effect on action. They are the actions we analyze, study and try to understand; and, starting from this, many scholars have proposed many emotion theories (e.g., top-down theories; versus bottom-up theories that analyze the biological basis of emotions).

In non-temporal terms, we believe, as stated earlier, that it is useful to categorize emotions as follows:

- emotions linked to the survival of individuals and the species;
- emotions not linked to the survival of individuals or the species.

Emotions linked to the survival of individuals and the species could be inserted in artificial agents. With respect to human beings, examples of such emotions are fear and anger, strictly connected to defense and attack. In terms of artificial agents, with the tools and knowledge available today, it is possible to consider defense and attack strategies utilizing the same general effect generated by emotions.

To give an example, we can consider the stock exchange and its fluctuations; considering particularly the simulations performed by Ferraris-Cappellini and Cipresso-Villamira, and submitted to WIVACE 2007. In the first simulation (FC), the agents are endowed with memory features (i.e., the agents remember less the information received – temporally – in central periods, memory follows an inverted Gaussian curve, so agents remember the tails more: primacy and recency effects). The agents are able to transfer information to each other to achieve operative consequences. In the second simulation (CV), the stock markets are analyzed in hyperinflation situations. Let us conjecture that an announcement is made that the supply of money is going to be increased at some time in the future. In actuality, the Central Bank has changed nothing as of the moment of the announcement. Nonetheless, prices change almost immediately, probably because of changes in the agents' perceptions with respect to monetary policy; which, in turn, alter their expectations regarding price levels.

Obviously, even without considering any emotions in the agents, both simulations are adaptable to a model like this. Defense and attack mechanisms among agents, on the basis of information on the market, could well represent an emotional framework even with selection mechanisms.

2.1. Emotions and artificial agents

Including emotions does not imply getting agents to make decisions quickly. If an agent encounters an obstacle, that agent 'decides' which direction to take on the basis of learning and a variety of emotional issues.

To consider emotions at the agent level, we need, first of all, to provide an emotional framework at this same level.

At the computational level, this may be realized with models that incorporate the following categories of agent:

- *Risk-adverse agent* (emotional feature: fear): if it prefers to obtain certainty in the expected value of a certain aleatory quantity instead of the aleatory quantity itself;
- *Neutral agent* (no emotion): if an agent is indifferent to the choice between the expected value of a certain aleatory quantity and the quantity itself;
- *Risk-inclined agent* (emotional feature: anger/aggressiveness): if the agent always prefers to obtain a certain aleatory quantity, instead of its expected value.

This categorization scheme is neither obviously exhaustive, nor perfectly exact, and surely may be linked to many other variables. However, we need it to create an 'easy artificial framework' that discriminates agents at the various extremes: e.g., the most risk-adverse agent versus the most risk-inclined agent.

The transitions 'from defense to fear' and 'from attack to anger' are forced for our needs (or better, for the needs of the model/simulation).

2.2. Choice/behavior processes of artificial agents

Within the choice/behavior processes of artificial agents, we consider environmental elements (and not only these) that exert their effect through emotional processes; and these can – for instance – act following the attack and defense paradigms, thereby affecting other agents who may adopt similar mechanisms.

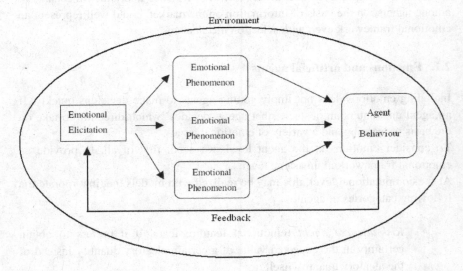

Figure 3: From the elicitation of emotions are emotional phenomena derived, on the basis of the hazard adversity/neutrality/propensity of agents, which implies behaviors that will generate new stimuli/emotions/responses, etc.

References

Ciarrochi J., Mayer J.D. (2007), Applying Emotional Intelligence, Psychology Press, New York

Cappellini A.N., Ferraris G. (2007) Evoluzione culturale in un sistema finanziario attraverso un algoritmo pseudo memetico in Proc. of WIVACE 2007

Cipresso P., Villamira M.A. (2007), Aspettative razionali e microfondazione per ecomomie in situazione di iperinflazione in Proc. of WIVACE 2007

Dawkins R. (1989), The selfish gene, Oxford University Press, Oxford

Frijda N.H. (2008), The laws of emotion, Psychology Press, New York

Kahneman D., Tversky, A. (1979), "Prospect Theory: An Analysis of Decision under Risk", Econometrica, vol. 47, no. 2

LeDoux J. (1998), Fear and the brain: Where have we been, and where are we going?, Biological-Psychiatry, 44, 12

Lewis M, Haviland-Jones J.M. & L. Feldman Barrett (Eds.), (2008), Handbook of Emotions. Third Edition, Psychology Press, New York

Minsky M. (1986), The Society of Mind, Simon and Schuster, New York

Pfeifer R. (1988), Artifical Intelligence Models of Emotion. In V. Hamilton, G.H. Bower, & N.H. Frijda (Eds.), Cognitive Perspectives on Emotion and Motivation, Kluwer, Netherlands

Plutchik R. (1980), Emotions: a psychoevolutionary synthesis, Harper & Row, New York

Plutchik R. (2002), Emotions and Life: Perspectives for Psychology, Biology, and Evolution, American Psychological Association, Washington, D.C.,

Power M., Dalgleish T. (2008), Cognition and emotion. From order to disorder, Psychology Press, New York

Simon H. (1997), Models of Bounded Rationality, Vol. 3. MIT Press

Velásquez, J. (1997). Modeling Emotions and Other Motivations in Synthetic Agents. In Proceedings of the AAAI Conference 1997, Providence

Villamira M.A. (in press) Comunicare, FrancoAngeli, Milano

EVOLUTION OF HIGH LEVEL RECURSIVE THINKING IN A COLLISION AVOIDING AGENT MODEL

F. ZANLUNGO

Department of Physics, Università di Bologna,
via Irnerio 46, 40126, Italy
E-mail: zanlungo@bo.infn.it

We introduce a collision avoiding method for simulated agents based on recursive thinking, in order to understand if more developed "Theory of Mind" abilities (used to predict the motion of the others) can allow the agents to perform in a better way in a complex environment. Agents move in their environment trying to reach a goal while avoiding collisions with other agents. They try to predict the motion of the others taking in consideration also the others' prediction. This introduces a recursive process, that we stop at a given "recursion level". We study the evolution of this level, showing that there is a difference between even and odd levels, that leads to a correspondence with the hawk-dove game, and that the collision avoiding abilities grow with the recursion level l, at least when the problem to be solved is complex enough. We also show that if the fitness function is properly chosen, the system converges to a stable state composed almost only of high level agents.

Keywords: Theory of Mind; Evolutionary Game Theory; Multi Agent Models

1. Introduction

Human beings have developed the ability to assume the "Intentional Stance"[1] when dealing with other humans (but also with animals and complex artifacts), i.e. they try to predict the behaviour of an opponent assuming that she has intentions and beliefs. Following the seminal work by Premack,[2] an individual with this ability is said to have a Theory of Mind (ToM). Human Theory of Mind can assume that also the others have a ToM, and thus is capable of nested mental states or higher order ToM,[3] leading to some kind of recursive structure (I think that you believe that she thinks...).

A large amount of research has been dedicated to understanding if non-human primates have a ToM (or to what extent they have it),[4] and to understand the relation between Theory of Mind deficits and Autistic Spectrum Disorders.[3] According to the "Machiavellian Intelligence" hypothesis,[5] high cognitive properties, including ToM, evolved in primates as the result of strong social competition, since they allowed for an higher (social and thus mating) success. In this work we propose an evolutionary computation model with agents capable of different levels of "recursive thinking" (or "nested mental states"). The situation that our agents have to face (moving in a crowd) is not complex enough to necessarily require high ToM levels,[6,7] but allows for a sound description in a realistic even if simulated physical space. Our agents are 2D discs, whose physical dynamics is exactly resolved, each one provided with a spatial goal (a region it wants to reach). Following Takano et al.[8,9] we call level 0 an agent that moves straight to the goal, without taking into account the others, while a level 1 agent can observe a neighbouring region of space and predict its physical dynamics in order to regulate its motion. Nevertheless a level 1 agent has no ToM, i.e. it assumes that the dynamics of the others is purely physical (it assumes them to be level 0). A level 2 agent is capable of "first order" ToM, i.e. it assumes that also the other agents have a physical model of their environment (assuming them to be level 1). "Second order" ToM[10] is attained by level 3 agents assuming that the others are level 2, and so on.

2. The Model

Our model is based on simulated agents, each one being a 2D disc with radius R moving in a corridor, i.e. in a 2D space delimited by two parallel walls and with two open ends. A "crowd" of $N \approx 10^2$ agents is roughly uniformly divided in two groups, each one with a different goal (one of the ends of the corridor).

The dynamics of the system can be divided in a physical and a "cognitive" one (i.e., a force than depends on the perception and on the internal properties of the agent). The latter, described in detail later, is applied simultaneously by all agents at discrete times with time step Δt, and acts as an impulsive force $\mathbf{f}_c$ according to

$$\mathbf{v}(t) = \mathbf{v}(t - \Delta t) + \mathbf{f}_c(t)\,\Delta t \tag{1}$$

$$\mathbf{x}(t + \Delta t) = \mathbf{x}(t) + \mathbf{v}(t)\,\Delta t \tag{2}$$

If the magnitude of the velocity is greater than a maximum value v_{max}, the velocity is simply scaled to v_{max} while preserving its direction. The physical dynamics is given by elastic collisions between the discs and with the walls, and is exactly solved as a continuous time function using an event driven algorithm. The cognitive force $\mathbf{f}_c$ is the sum of an external term $\mathbf{E}$ (a constant field directed towards the agent's goal) and a collision avoiding term $\tilde{\mathbf{f}}_{\mathrm{int}}$

$$\mathbf{f}_c = \mathbf{f}_{\mathrm{int}} + \mathbf{E} \qquad (3)$$

$\mathbf{f}_{\mathrm{int}}$ depends on the interaction with the other agents and is determined by the agent's "level of recursive thinking" l. By definition, $\mathbf{f}_{\mathrm{int}} = 0$ for a $l = 0$ agent, i.e. level 0 agents have no cognitive interactions. $l > 0$ agents observe the position and velocity of all the other agents that are located at a distance $d < r_v$ (the "radius of view" of the agent). On the basis of this observation they forecast the future evolution (of the physical and cognitive dynamics) of this portion of the system for a time $t_f = n\Delta t$, with n a positive integer. While the physical dynamics is trivially defined by elastic collisions between agents and with the walls, the cognitive one is recursively determined assuming that all the observed agents will move as $l-1$ agents (where l is the level of the agent performing the prediction) while the agent itself will move as a $l = 0$ one (this definition can be explained in the following way: $l - 1$, by definition, is the highest level of prediction at which a level l agent can forecast the others' motion, while in predicting its motion the agent assumes that it will move straight to its goal, in order to attain the highest performance). While forecasting the evolution of the system, the agent keeps track of all the (predicted) collisions that it will have with the others and with the walls. Denoting t_i as the time of the i-th predicted collision, and $\mathbf{p}_i$ as the total momentum exchanged during the collision (see Fig. 1 for a definition and explanation of $\mathbf{p}$), we have

$$\mathbf{f}_{\mathrm{int}} = \sum_i \frac{\mathbf{p}_i}{t_i} \qquad (4)$$

(The agent changes its velocity in order to avoid the collision, assuming that the others will keep their velocity according to the prediction. The idea at the basis of this definition is to avoid strongly any predicted collision, so that the prediction abilities of our agents can be reliably measured by the amount of collision. Nevertheless other methods can lead to more smooth trajectories[6,7] and to the emergence of self organised patterns as those shown by actual pedestrians[11]).

All the operations concerning both the observation of another agent's position and the prediction of its motion are subject to a random (relative) error of order 10^{-4}.

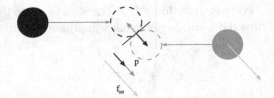

Fig. 1. Definition of **p** in Eq. (4). The green (light grey in b&w) ball is forecasting its motion and that of the blue (dark grey) one. At the moment of collision $\bar{t}$ (dashed balls) it subtracts the component of its momentum orthogonal to the surface of collision to the component of the momentum of the blue ball in the same direction. The result is **p** which is then scaled as $\mathbf{p}/\bar{t}$ to obtain the force $\mathbf{f}_{\text{int}}$ which acts on the motion of the green ball. In case of a collision with a wall we have the same procedure but the term due to the other agent (blue arrow) is set to 0. In the latter case **p** is just half exchanged momentum.

2.1. *Experimental Setup*

The agents have radius $R = 0.5$ m, are located in a corridor of length 50 m and width 5 m and can move with a maximum velocity $v_{\max} = 1\,\text{m/s}$, while the attraction to the goal is $E = 0.5\,\text{m/s}^2$ (masses are considered fixed to 1, and thus accelerations equal to forces) and the radius of view is $r_v = 3$ m. We use as time step $\Delta t = 0.4$ s while the time step of prediction of the future dynamics by the agents is set to $t_f = 3\,\Delta t = 1.2$ s. We use a population of $N = 50$ agents, with values of l in the 0-4 range, randomly given in the first generation using a uniform probability distribution.

In the evolutionary experiments each generation consists of three tests, each one of time length $T = 100$ s. In each test agents are initially randomly located in the corridor, and when they reach their goal (i.e., a given end of the corridor) are relocated at the other end (in a randomised transverse position). During each generation and for each agent we keep track of its average velocity towards the goal ($\bar{v}$) and average momentum exchanged in collisions ($\bar{p}$), and we evaluate its individual performance using a fitness function given by

$$f = \bar{v} - \beta\,\bar{p} \tag{5}$$

where $\beta \geq 0$ is a parameter that determines the relative weight of collision with respect to velocity. We use this fitness function because we want agents to be able of moving in the direction of the goal, while avoiding collisions. We underline that the collision avoiding ability that we have introduced in our model has been thought in order to minimise $\bar{p}$, which is a measure of the prediction ability of the agents. Nevertheless, using fitness (5) we introduce the benefit of exploiting the collision avoiding abilities of another agent (as we show below), and thus a more interesting evolutionary dynamics.

The genetic algorithm uses tournament selection (two agents are randomly chosen in the previous generation, their fitness is compared and the winner passes its character -i.e., its value of l- to the next generation). The mutation operator acts with probability $p_m = 0.05$, changing l to a different value randomly chosen in the allowed range.

3. Behaviour for Different Values of l

Before performing any evolutionary experiment we have analysed the collisional properties of our agents under controlled conditions, in order to understand more deeply some features of the model.

First of all we have studied the behaviour of agents in binary collisions. To do that we have repeated 1000 times an experiment in which two agents with different goals and colliding trajectories were located in a corridor without walls at a distance comparable to their average free walk under our experimental conditions. The $\bar{v}$ and $\bar{p}$ values attained in these experiments, for collisions between agents in all the possible combination of l in the 0-4 range, resulted to be modulus 2 symmetric, i.e. even (or odd) levels are not distinguishable between them (in binary encounters) and thus the system can be completely described by a 2×2 f (fitness) matrix (Table 1) that, for any value of β, takes the form of the classical hawk-dove game matrix,[12] where even levels correspond to hawks, odd to doves (an asymmetry between even and odd levels had been already found by Takano et al.[8,9]). The reason of this symmetry can be understood noticing that we

Table 1. Fitness in binary encounters. f_{ij} gives the fitness attained by i in an encounter with j.

l	even	odd
even	0.56 - 0.27 β	1
odd	0.64	0.76

never had, in all our simulations of binary encounters, a collision concerning an odd level agent (this is the reason β appears only in the even-even term of f_{ij}). When interacting with a $l = 0$ agent, a level 1 agent predicts in an accurate way the motion of its opponent, and avoids the collision. Since a $l = 2$ agent, interacting with any other agent, will predict the motion of this one as if it were level 1, and its motion as if itself were level 0, it will predict no collision ($f_{\text{int}} = 0$, Eq. (4)), and thus $l = 2$ is completely equivalent to $l = 0$ (and by iteration we obtain the symmetry between all even and all odd levels).

Following[12] we can calculate an evolutionary stable state from Table 1 as the portion of agents in odd levels $x_o \equiv N_o/N$ (where N_o is the number of odd level agents) for which the average odd fitness is equal to the average even one, which is

$$x_o = \frac{f_{oe} - f_{ee}}{f_{oe} + f_{eo} - f_{oo} - f_{ee}} = \frac{0.64 - (0.56 - 0.27\beta)}{0.64 + 1 - 0.76 - (0.56 - 0.27\beta)} = \frac{0.08 + 0.27\beta}{0.32 + 0.27\beta} \tag{6}$$

(f_{oe} is the fitness of an odd agent when meeting a even one, and so on, see Table 1). Equation (6) gives some qualitative results for the evolution of a mixed population of level 0 and level 1 agents (the system seems to converge to a stable value of x_o, and this value grows with β, see Fig. 2, left) but the results are not in quantitative agreement with the predicted ones. Furthermore, the evolution of a level 1 and 2 population still converges to a stable x_o that grows with β, but at fixed β these values are not equal to those obtained in the level 0 and 1 case (Fig. 2, right). These

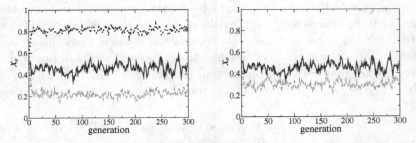

Fig. 2. Left: evolution of the number of odd level agents x_o in a population with $l = 1$ and $l = 0$ agents ($N = N_e + N_o = 50$), for $\beta = 0.5$ (green, light grey in b&w-dashed), $\beta = 1$ (blue) and $\beta = 2$ (black-dotted). Right: evolution of x_o with $\beta = 1$ in a population composed of $l = 0, 1$ (blue) and $l = 1, 2$ (green-dashed) $N = 50$ agents.

results show that the multi-agent dynamics have properties that cannot be analysed just studying binary encounters. Table 2 reports the f matrix for encounters concerning 4 agents, two from each species. The interaction with more than a single agent makes impossible for $l = 1$ agents to avoid all the collisions, and thus breaks the symmetry between $l = 0$ and $l = 2$ (the term in Eq. (4) is now different from zero for $l = 2$, thus these agents interact and usually behave better, in particular concerning $\overline{p}$, than $l = 0$ agents). These results can explain those in Fig. 2: first of all, they show that

Table 2. Fitness in encounters with 4 agents. f_{ij} gives the average value attained when 2 agents of level i meet 2 agents in level j.

l	0	1	2
0	0.40 - 0.61 β	0.75 - 0.21 β	0.42 - 0.57 β
1	0.32 - 0.04 β	0.62 - 0.01 β	0.35 - 0.01 β
2	0.42 - 0.50 β	0.81 - 0.10 β	0.43 - 0.43 β

it is impossible to obtain quantitative estimates of the fitness of agents in a crowd just from binary encounters (since our dynamics does not follow a superposition principle); second, that it is plausible that level 2 agents will behave better than level 0 ones when interacting with level 1 (in particular for high values of β). From the analysis of Table 2 we observe that, even if the even-hawk, odd-dove analogy is still valid (even level have the tendency to move straight, while odd ones to avoid the collision), the matrix corresponds to the classical one only for a given range of β.

Nevertheless, Table 2 too does not provide quantitative information about the dynamics of a large population, in which the presence of limited knowledge effects causes a larger number of level 1 collisions and thus a more complex l dependence of the dynamics. Figure 3 shows $\overline{v}$ and $\overline{p}$ in a homogeneous population of 50 agents (i.e, all agents have the same value of l), as a function of l. We can see the difference between even and odd levels, but also a tendency to increase $\overline{v}$ and decrease $\overline{p}$ as l grows, both for even and odd levels (the only exception being $\overline{v}$ for even levels, which is almost constant). Notice that, in comparison with the results on the diagonal of Table 2, the amount of $l = 1$ collisions has increased by an order of magnitude and that, correspondingly, also the difference between $l = 0$ and $l = 2$ has grown.

4. Evolutionary Experiments

The fitness of our agents is determined by a function $f = f_l(\{x_l\}, \beta)$ of β, of their level l and of the composition of the population $\{x_l\}$, $x_l \equiv N_l/N$. If we knew this function, we could study the population dynamics using a replicator equation.[12] As we have seen, this dependence cannot be derived analysing encounters between a low number of agents, but just studying the actual dynamics of a crowd of agents (only when N is large enough

162

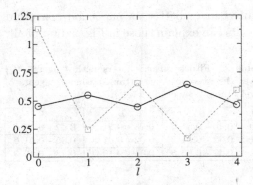

Fig. 3. $\bar{v}$ (black, continuous, circles) and $\bar{p}$ (green, dashed, squares) as a function of l. These data were obtained as averages over 100 tests with $N = 50$ agents, each test lasting $T = 100\,s$.

the number of $l = 1$ collisions becomes significant). Even if the dependence on β is trivial (because this parameter has no effect on dynamics, only on evolution), since we have verified that $f_l(\{x_l\}, \beta)$ is not linear in x_l, an estimate of this function good enough to give quantitative results for l in the 0-4 range would require a large number of tests in order to obtain a data set suitable to some kind of interpolation. This kind of analysis would be affordable from a numerical point of view, but scarcely meaningful since its computational cost is much higher than that of a full evolutionary experiment, and thus we have not performed it at this stage.

The results of our evolutionary experiments (averages over 10 repetitions of the experiment) are shown in Fig. 4. We can see that in the $\beta \to 0$ limit the population is invaded by level 0, in the $\beta \gg 1$ limit by level 3, while for $\beta \approx 1$ the population is almost uniformly split in the 3 higher levels (the system seems to converge always to a stable state).

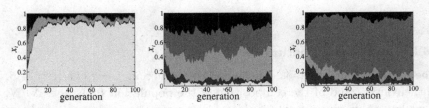

Fig. 4. Evolution of x_l ($l = 0$ in yellow, $l = 1$ in blue, $l = 2$ in green, $l = 3$ in red, $l = 4$ in black. The different areas corresponding to each level are ordered in growing values of l from bottom, i.e. $l = 0$, to top, i.e. $l = 4$). Data obtained as averages over 10 repetitions with $N = 50$ agents. Left $\beta = 0$. Centre $\beta = 1$. Right $\beta = 10$.

5. Analysis

While the $\bar{p}$ term in Eq. 5 measures directly the collision avoiding properties of agents and is symmetric, i.e. it has the same value for all the agents involved in a interaction (at least in binary collisions), $\bar{v}$ depends only indirectly on collision avoiding and can be highly asymmetric, allowing the exploitation of another agent's behaviour. For this reason the value of β determines the degree of cooperation of the system, leading to selfish behaviour if $\beta \to 0$ and to cooperative behaviour if $\beta \to \infty$. In the $\beta \gg 1$ regime our system is invaded by the highest possible odd level, showing that this level has the highest ability to predict the evolution of the system and thus to avoid collisions. Lowering the value of β the number of even agents at equilibrium increases, in qualitative agreement with equation 6. Nevertheless, while for $\beta \approx 1$ the even population is invaded by high levels, due to their higher prediction and thus collision avoiding properties, when β goes to 0 the completely "selfish" $l = 0$ agents invade the population. $l = 1$ agents never have a major role in the process, showing that their behaviour differs from $l = 3$ only in a lower prediction ability.

The results of the evolutionary experiments show that when a large number of agents are interacting, and thus the task of predicting the dynamics of the system is not trivial, the "modulus 2" symmetry is broken and to each value of l corresponds a different dynamics. In particular, the fact that for high values of β the population is invaded by agents with high values of l, shows that high recursion levels have better prediction properties.

6. Conclusions

We have seen that a collision avoiding system based on high level recursive thinking prediction of the motion of the others is both more effective and (at least if the fitness function is chosen in such a way to allow the evolution of a cooperative behaviour) favoured by the evolutionary process. In this first model we have not considered the cost of computation, which grows exponentially with l. Since we have noticed that the difference between low and high levels grows with the difficulty of the problem (in this case the number of $l = 1$ collisions) it is probable that the balance between the cost of computation and the benefit due to more precise prediction is determined by the nature of the problem (we intend to study this aspect more in depth in a future work).

In this model we have allowed our agents to have a 360 degrees vision, limiting the effects due to partial knowledge that are surely interesting in a recursive thinking model and that will be analysed in future.

This work presents also the difference between even and odd levels discovered by Takano et al.[8,9] It can seem strange that the evolution of recursive thinking leads to such a discontinuous behaviour, but we stress that in our work, as in Takano's, we are considering just *pure strategies*. Probably an actual level l agent should not just consider the results of its level l calculations, but would weigh them with those resulting from calculations at lower levels. The introduction of such weights in the "genetic code" of the agent should enhance strongly its computational power. Furthermore, these weights could be updated after each interaction using some reinforcement learning method, introducing a more realistic learning process and allowing the study of the evolution-learning interaction. We intend to study such a model in a future work.

References

1. D. Dennet, *The Intentional Stance*, MIT Press (1987).
2. D. Premack and G. Woodruff, *Does the Chimpanzee have a Theory of Mind?*, The Behavioral and Brain Sciences, 1:515-523 (1978).
3. S. Baron-Cohen, *Theory of mind and autism: a review*, Special Issue of the International Review of Mental Retardation, 23, 169, (2001).
4. J. Call and M. Tomasello, *Does the chimpanzee have a theory of mind? 30 years later*, Trends in Cognitive Sciences, Vol. 12 No. 5
5. R.W. Byrne and A. Whiten (Eds.), *Machiavellian Intelligence: Social Expertise and the Evolution of Intellect in Monkeys, Apes, and Humans*, Oxford Science Publications (1989).
6. F. Zanlungo, *Microscopic Dynamics of Artificial Life Systems*, Ph.D. Thesis, University of Bologna, (2007, unpublished).
7. F. Zanlungo, *A Collision Avoiding Mechanism Based on a Theory of Mind*, Advances in Complex Systems, Special Issue Vol.10 Suppl. No. 2.
8. M. Takano, M. Katō and T. Arita, *A Constructive Approach to the Evolution of the Recursion Level in the Theory of Mind*, Cognitive Studies, 12(3), 221-233, (2005, in Japanese).
9. M. Takano and T. Arita, *Asymmetry between Even and Odd Levels of Recursion in a Theory of Mind*, Proc. of ALIFE X, 405-411, (2006).
10. J. Perner and H. Wimmer, *"John thinks that Mary thinks that..."*: Attribution of second order false beliefs by 5 to 10-year-old children, Journal of Experimental Child Psychology, 39, 437-471 (1985)
11. K. Andō, H. Ōto and T. Aoki, *Forecasting the flow of people*, Railway Res. Rev. 45(8) 8-13, (1988, in Japanese).
12. J. Hofbauer and K. Sigmund, *Evolutionary Games and Population Dynamics*, Cambridge University Press (1998).

PART IV

ROBOTICS

WHO IS THE LEADER? DYNAMIC ROLE ALLOCATION THROUGH COMMUNICATION IN A POPULATION OF HOMOGENEOUS ROBOTS

ONOFRIO GIGLIOTTA, MARCO MIROLLI, STEFANO NOLFI

Institute of Cognitive Sciences and Technologies, CNR, via S. Martino della Battaglia 44
Rome, 00185, Italy

The field of collective robotics has been raising increasing interest in the last few years. In the vast majority of works devoted to collective robotics robots play all the same function, while less attention has been paid to groups of robots with different roles (teams). In this paper we evolve a population of homogeneous robots for dynamically allocating roles through bodily and communicative interactions. Evolved solutions are not only able to efficiently decide who is the leader, but are also robust to changes in team's size, demonstrating the ability of Evolutionary Robotics to find efficient and robust solution to difficult design challenges by relying on self-organizing principles. Our evolved system might be used for improving robots' performance in all the cases in which robots have to accomplish collective tasks for which the presence of a leader might be useful.

1. Introduction

In the recent years the study of collective robotics has been raising increasing interest within the Artificial Life and Adaptive Behavior communities. In particular, Evolutionary Robotics techniques [13], [8] have been successfully used for designing robot controllers able to display swarm behaviors, that is behaviors in which a group of robot appears to act as a single unit (see, for example, [5], [3]). Evolutionary Robotics seems to be particularly well suited for designing such kind of robots. One of the main advantages of Evolutionary Robotics is in fact the ability of artificial evolution to find interesting solutions to difficult robotic tasks by exploiting the self-organizing properties of the complex dynamics of the interactions between a robot's control system, its own body, and the environment [12]. When dealing with groups of robots the complexity of the resulting system increases, since the interactions between the robots add up to the interactions between single robots and their environment so to produce an extremely complex and highly unpredictable dynamical system. Such systems are known to be very difficult to engineer by direct design, and this is the reason why Evolutionary Robotics has been raising increasing attention in the field of collective robotics.

Typically, evolved groups of robots constitute swarms, that is groups in which each individual behaves according to the same simple rules, so that the complexity of the behavior of the group emerges from the local interactions between the members. Though extremely interesting, this kind of organization does not permit to develop more complex social behaviors requiring specialization within the group. The reason of this is that typically interesting collective tasks require cooperativeness between the robots and this is typically assured in Evolutionary Robotics experiments by using groups of homogeneous robots. In fact, if interacting agents are non-homogeneous, then the problem of altruism immediately arises, making the emergence of cooperative behaviors extremely difficult (two examples of works devoted to the problem of altruism in groups of communicating agents are [11] and [6]).

It would be extremely useful to exploit Evolutionary Robotics techniques in order to develop teams of robots, that is groups of robots in which (1) different individuals make different contributions to the success of the task, (2) roles are interdependent thus requiring cooperation, and (3) organization persists over time (cf. the definition of 'team' provided by [1], in the contest of animal behavior). This poses a difficult but interesting challenge since it is not clear how homogeneous individuals might be able to assume different roles in a persistent manner. A possible solution to this problem might come from endowing robots with communication capabilities, so that role allocation might be negotiated through the exchange of signals. In the recent years several interesting studies have demonstrated the possibility to evolve communication in homogeneous robots so to accomplish some cooperative task (e.g. [4], [14], [9]). If we can evolve groups of homogeneous robots which are able to negotiate their roles through the exchange of signals, then this ability might be later exploited for the evolution of more complex collective behaviors requiring role specialization.

In this paper we describe an experiment in which a group of homogeneous robots is evolved for the ability to negotiate their roles: in particular, one of the robots has to maximize its communicative signal, thus assuming the role of the group's leader, while all the other robots have to minimize their signal. We show that evolved robots are not only able to solve the task, but that evolved solutions are also robust with respect to the number of interacting robots. In the next section we briefly review the two works which are most related to the present one. In section 3 we present the experimental set-up, while in section 4 we show the main results. Finally, section 5 concludes the paper with a brief discussion about the significance of this work and about possible lines of future work.

2. Related Work

To the best of our knowledge, there are only two published works devoted to the study of dynamic role allocation within an Evolutionary Robotics framework: [2] and [15].

Baldassarre and colleagues [2] evolved a group of robots for the ability to aggregate and collectively navigate toward a light target. Apart from infrared and ambient light sensors, robots were equipped with directional microphones, which could detect the sounds emitted by other robots' speakers with a fixed amplitude and a frequency randomly varying within a limited range. While analyzing the various results of different replications of several evolutionary runs, they found that three different kinds of strategies were discovered, which they called 'flock', 'amoeba', and 'rose'. What is most interesting for the purposes of the present paper is that the most efficient solution, the flock one, required different individuals playing different functions, with the individuals which are nearest to the light leading the group straight toward the target at maximum speed and the other individuals following the leaders by trying to maintain the group's cohesion. Since the groups were formed, as usual, by homogeneous individuals, and since robots' controllers were formed by simple perceptrons and hence did not have any internal state, robot's specialization was 'situated', in the sense that it completely depended on the different input patterns that robots received from the environment.

In a similar work, Quinn and colleagues [15] evolved a team of three uniform robots able to dynamically allocating their roles in order to navigate as a group. In this experiment, robots equipment was really minimal: each robot had just four infrared sensors and two motor-driven wheels. Robots were required to move as a group, that is by remaining within each other's sensor range. The analysis of evolved robots' behavior showed that the task was completed by relying on two phases: during the first phase robots organize themselves into a line formation, while in the second phase the robots start to move swinging clockwise and anticlockwise while maintaining their relative positions.

While the behaviors of the robots evolved by Baldassarre and colleagues and Quinn and colleagues have several interesting features, the solutions found by evolution are not general, but rather task-specific: the dynamic role allocation performed by those robots can be used only for the specific collective navigation tasks the robots were evolved for, and cannot be exploited for other kind of purposes. Furthermore, that particular solutions do not seem to have the robustness which is typically assured by the use of homogeneous teams. As correctly indicated by Quinn and colleagues, while the predetermination of roles has the clear advantage of not requiring on-line role allocation and of permitting robots' behavioral and morphological specialization, on the other hand the use of homogeneous robots has the potential of being much more robust: since all robots are identical and hence each robot is equally able to play any role, teams of homogeneous robots can potentially be much more able to cope with the loss of individual members with respect to heterogeneous teams. This is clearly true, but this advantage is not demonstrated neither in the work of Baldassarre and colleagues nor in that of Quinn and colleagues. In fact, Baldassarre et al. did not touch the problem of robustness at all, and it is not clear whether the flocking behavior would generalize with respect to the number of robots. On the contrary, the solution found by the robots of Quinn and colleagues did demonstrate *not* to be robust to the lack of an individual: if one of the three robots is removed from the formation the remaining pair maintain the same configuration as when in full formation but ceasing the forward movement. The reason is that the evolution of task-specific role allocation will tend to produce task-specific solution which deeply rely on the specific conditions under which evolution took place.

A possible solution to this problem might consist in directly evolving groups composed by different numbers of robots to perform the same task. In this paper we explore another, much more general, solution: namely, the direct evolution of the ability to dynamically allocating roles between themselves through the use of local communicative signals. Once we have reached a group of robots which are able to negotiate their roles on the fly, we might exploit this ability for solving, with a population of homogeneous robots, any kind of collective robotic tasks requiring role differentiation.

3. Experimental Setup

The experimental setup consists in a group of four identical e-puck robots (Figure 1 left) placed in a box-shaped arena of 40x40 cm (Figure 1 right).

Figure 1. Left: E-puck robot. Right: a group of four e-puck robots inside a box shaped arena of 40x40cm

Robots can move in the arena by sending commands to their two wheels and can exchange signals between each others through a dedicated communication channel. Communication is local as each robot perceives only the signal which is emitted by its nearest fellow. Robots' signal are not only used for communication: they also represent the role of the signaling robot. We evolve the robots for the ability to differentiate their roles through the differentiation of their signals: one of the robots must become the 'leader' of the group by maximizing the value of its communicative output, while all other robots, which are non-leaders, must minimize the values of their signals. More concretely, we calculate the fitness of a group of robot in the following way. For each cycle, we consider the average of the differences between the communicative output of the current 'leader' (i.e. the robot with maximal communicative output) and the communicative outputs of all other robots. The fitness is the average of this value for all the cycles of all the trials. Formally, this is how fitness is calculated:

$$F = \frac{\sum_j^C \sum_i^N Max_j - O_{ji}}{C(N-1)} \qquad (1)$$

where N is the number of robots in the group (i.e. 4), C is the total number of life-cycles of each individual (i.e. 1000 cycles times 40 trials = 40000), MAX_j is the value of the signal of the leader at cycle j and O_{ji} is the value of the signal of robot i at cycle j.

172

Each robot is controlled by a neural network with a fixed architecture shown in Figure 2. There are ten sensory neurons: 8 input units encode the state of the 8 infrared sensors, which are disposed around the robot's body; one input unit encodes the signal emitted by the nearest robot, and the last input unit encodes the activation of the same robot's communicative output units during the previous cycle. All the input units send connections to two hidden units, which are leaky integrators with fixed time constants (set to 0.9) and are fully connected between themselves. Finally, the two hidden neurons send connections to both the communicative output unit and to the two motor output units, which are used to command the motors of the two wheels and receive also direct connections from the 8 infrared input units.

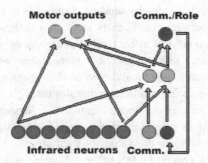

Figure 2. Neural controller

The free parameters of the robots' neural controllers are evolved as in a standard Evolutionary Robotic setup [13]. The genome of individuals encode all the network's connection weights and biases. Each parameter is encoded as an 8 bits string, whose value is then uniformly projected in the range [−5.0, +5.0]. The initial population consists of 100 randomly generated genotypes. Each genotype is tested for 40 trials, lasting 1000 cycles each. At the beginning of each trial, four homogeneous robots (i.e. with the same connection weights) are randomly placed into the arena, and they are left free to move and to communicate between each other for all the rest of the trail. After all the trials of all individuals within a generation have been executed, the 20 best genotypes of each generation are allowed to reproduce by generating five copies each, with 2% of their bits replaced with a new randomly selected value. The evolutionary process lasts 150 generations (i.e. the process of testing, selecting and reproducing robots is iterated 150 times). All the experiments were carried out in simulation using the Evorobot simulator [13] adapted so to simulate the e-puck robots.

4. Results

The experiment was replicated 10 times, with different random initial conditions. The average fitness of the best individuals (teams) of all the replications is about 0.82, with the best replication achieving a fitness of about 0.9. By looking at the behaviors exhibited by the best groups of the last generations of all the runs, we could see that in 9 out of 10 replications robots are able, after a short transient during about 100 cycles, to dynamically allocate their role in an efficient and quite stable way. Only in the worst run, which achieved a best fitness of about 0.62, roles never stabilize, with every robot continually changing its communicative behavior in an apparently chaotic way. With respect to the 9 successful replications, we noted that different replications found slightly different solutions, but the behaviors of the best individuals are qualitatively quite similar.

The robots within a group tend to differentiate quite rapidly, with one robot assuming the leader role (signaling with a very high value) and the other three robots setting their communicative output signal at very low values. After differentiation, robots remain in their role for almost the whole life time, although small changes can occur due to the intrinsic dynamical negotiation process. In fact, due to the local communication system, in order to hold a leader role, a robot have to continuously interact with the non-leader robots. This has the interesting consequence that not only the communicative behavior is differentiated, but also the non-communicative one. In other words, although this was not explicitly requested in the fitness function, after role allocation even a robot's non-communicative behavior significantly depends on the role that robot plays within a group. In particular, leaders tend to have a broader explorative behaviour than non-leaders, which "prefer", instead, to reduce their movements. This is done in order to maximize the number of interactions that a leader can carry out with all other individuals of a group so to maintain its leadership. This kind of group organization is present in all the successful replications, which differ mainly in the kind of behavior exhibited by leaders (in general it is a circular behavior) and in the amount of movement exhibited by non-leaders (which in some cases just circle around themselves without any displacement).

From the point of view of signals, robots tend to produce only two signals: the 'leader' signal L, with very high values (in the best replication this signal assume the value of about 0.96), and the 'non-leader' signal NL, with very low values (in the best replication this signal assume the value of about 0.1). At the beginning of a trial signals are typically produced randomly, while after a few interactions only one robot is able to maintain an L signal since an L produced by a speaker tends to induce the hearer to answer with a NL signal.

How robust is the behavior of the evolved robots? Are these solutions able to generalize to groups of robots composed by a number of individuals which is different from the one the robots have been evolved to cope with? In order to check this we tested the performance of the best individual of the last generation of the best replication of our experiment in five different conditions: in groups of 2, 3, 4, 5, and 6 robots (remember that evolution was run only with groups of 4 robots). For each test, we calculated not only the average fitness (calculated according to the formula 1, which is clearly independent on the number of robots), but also the average number of leaders in the group. In this case, we decided to count as a 'leader' every individual whose communicative output value is above 0.5. We monitor the number of leaders throughout the whole test (i.e. for each cycle of each trial), and we report the percentage of the cycles in which there are different numbers of leaders (from 0 to 6). The results, shown in Table 1, clearly demonstrate a remarkable generalization ability of the evolved solution. Evolved robots are in fact able to effectively allocating roles within the groups even in conditions which have never been experienced during evolution. Indeed, instead of decreasing, performance actually increases if the group is composed by a number of robots which is inferior to the one with which evolution took place. In fact, performance seems to be linearly dependent on the number of robots present in the environment: the lower, the better. This is clearly due to the kind of strategy used by evolved robots to solve the task: as this requires a leader to navigate through the environment so to communicate its presence to all other robots, the less crowded is the environment, the easier is for the leader robot to reach all the other robot so to maintain its leadership. On the contrary, as the number of robots increases, the environment gets more crowded, and the easier it is to get to situations in which there are either two leaders or none.

Table 1. Average percentage of leaders during a lifetime of 1000 times 40 trials, in groups of 2, 3, 4, 5, and 6 robots. The last row of the table indicates average fitness, calculated according to Eq. (1).

N. Leaders	2 Robots	3 Robots	4 Robots	5 Robots	6 Robots
0	0.92	0.95	1.05	6	10.78
1	99.08	95.26	88.62	75.73	71.00
2	0	3.79	10.33	17.34	17.22
3	-	0	0	0.92	0.98
4	-	-	0	0.01	0.02
5	-	-	-	0	0
6	-	-	-	-	0
Average fitness	0.95	0.92	0.90	0.83	0.81

5. Discussion and Conclusion

In this paper we presented evolutionary robotics experiments in which groups of four robots are evolved for their ability to dynamically allocate their roles through their communicative and non-communicative interactions. In particular, evolved robots are able to differentiate both their communicative and non-communicative behaviors so that only one robot assumes the role of the 'leader' of the group, sending high value signals, while all the other robots act as non-leader, almost ceasing their signaling behavior. The task is interesting because the groups of robots are homogeneous (i.e. all the group's members have the same body and control system): as a consequence, robots need to negotiate their roles on the fly. Furthermore, in contrast to most previous works dealing with dynamic role allocation, in which robots can rely on predetermined communication protocols by which robots can share global information (e.g. [10], [16]), in our experiments robots can rely only on the local information provided by their infrared sensors and by a one-to-one communication channel.

The most interesting result of our simulations is related to the generalization abilities of our evolved solutions. In contrast to the two previously published evolutionary robotics works dealing with dynamic role allocation [2], [15], which did not demonstrate any generalization ability, our system proved to be very robust to changes in the number of robots forming a group. While evolved for allocating the role of the leader in groups made up of four individuals, evolved solutions are able to perform reasonably well this kind of dynamic role allocation even in groups of five and six robots, while the efficiency of the evolved strategy is *increased* as the number of robots forming a group decreases (i.e. in groups of 2 and 3 robots). These results clearly demonstrate the feasibility and the potentiality of our Evolutionary Robotics approach to the development of complex collective (social) behaviors in autonomous robots. In fact, the flexibility and robustness demonstrated by our evolved solution has been possible only thanks to our use of groups of homogeneous robots, in which robots' roles are not pre-specified, but must be negotiated thanks to the dynamical interactions between the robots themselves.

We envisage at least three ways for extending the work presented in this paper. The first one, which we are currently exploring, consists in analyzing in more details the strategies of our evolved robots, in order to better understand how roles are dynamically allocated during the transient period at the beginning of each trial and how they are maintained once they have been decided.

A second interesting line for future research is related to the possibility of using our role allocation system as the starting point for developing robots able to accomplish collective tasks which require the presence of a leader. While several swarm-like behaviors might be successfully accomplished by groups of robots without any significant distinction between the behaviors of the members of the group, there are many cases in which the presence of a leader might significantly improve the performance of the group (see [1] for examples in the animal kingdom and [2] and [7] for examples within the artificial life community). In order to develop groups of homogeneous robots able to efficiently accomplish these kind of tasks, we might use our evolved robots, which are already able to rapidly negotiate who is the leader, for seeding the task-specific evolutionary search.

Finally, since the idea of using local communicative interactions between homogeneous groups of robots for dynamic role allocation is not strictly related to the development of a (single) leader, the same idea might be exploited also for developing robots able to dynamically allocate different kinds and numbers of roles.

References

1. C. Anderson and N. R. Franks, *Behavioral Ecology* **12**, 534 (2001).
2. G. Baldassarre, S. Nolfi, and D. Parisi, *Artificial Life* **9**, 255 (2003).
3. G. Baldassarre, D. Parisi, and S. Nolfi, *Artificial Life* **12**, 289 (2006).
4. E. A. Di Paolo, *Adaptive Behavior* **8**, 25 (2006).

5. M. Dorigo, V. Trianni, E. Sahin, R. Gross, T. Labella, G. Baldassarre, S. Nolfi, J. Deneubourg, F. Mondada, D. Floreano, and L. Gambardella, *Autonomous Robots* **17**, 223 (2004).
6. D. Floreano, S. Mitri, S. Magnenat, and L. Keller, *Current Biology* **17**, 514 (2007).
7. O. Gigliotta, O. Miglino, and D. Parisi, *The Journal of Artificial Societies and Social Simulation* **10** (2007).
8. I. Harvey, E. A. Di Paolo, M. Quinn, R. Wood, and E. Tuci, *Artificial Life* **11**, 79 (2005).
9. D. Marocco and S. Nolfi, *Connection Science* **19**, 53 (2007).
10. M. Mataric, *Adaptive Behavior* **4**, 51 (1995).
11. M. Mirolli and D. Parisi, *Connection Science* **17**, 325 (2005).
12. S. Nolfi, *Connection Science* **10**, 167 (1998).
13. S. Nolfi and D. Floreano, "Evolutionary Robotics: The Biology, Intelligence, and Technology of self-Organizing Machines." The Mit Press, Cambridge, 2000.
14. M. Quinn, *in* "Advances in Artificial Life : Sixth European Conference on Artificial Life" (J. Kelemen and P. Sosik, eds.), p. 357. Springer, Prague, Czech Republic, 2001.
15. M. Quinn, L. Smith, G. Mayley, and P. Husbands, *Philosofical Transactions of the Royal Society of London, Series A: Mathematical, Physical and engineering Sciences* **361**, 2321 (2003).
16. P. Stone and M. Veloso, *Artificial Intelligence* **110**, 241 (1999).

COOPERATION IN CORVIDS: A SIMULATIVE STUDY WITH EVOLVED ROBOT

ORAZIO MIGLINO, MICHELA PONTICORVO*, DAVIDE DONETTO

Natural and Artificial Cognition Laboratory, Department of Relational Sciences, University of Naples "Federico II" Naples, Italy
Institute of Cognitive Sciences and Technologies, CNR, Rome
* E-mail: michela.ponticorvo@unina.it
www.nac.unina.it

STEFANO NOLFI

Institute of Cognitive Sciences and Technologies, CNR, Rome

PAOLO ZUCCA

Department of Comparative Biomedical Sciences
Faculty of Veterinary Medicine, University of Teramo

1. Introduction

Corvids (*Corvidae*) family include various birds species characterized by high complexity in cognitive functions: they can be compared with primates both on brain relative dimensions, cognitive abilities and on social organization complexity (Emery, 2004; Emery and Clayton, 2004a, 2004b). They are capable of long term cache recovery, object permanence (Zucca et al., 2007), tool manipulation, theory of mind like-abilities (Bugnyar, 2007) and social reasoning. In nature we can observe them in dyads as well as in small or large colonies. Corvids are also able to cooperate in order to obtain a goal (Scheid and Noe, in preparation). Cooperation has attracted attention from many scholars coming from different disciplines such as psychology, sociology, anthropology, economics, that study human behaviour and ethology, behavioural ecology and evolutionary ecology that are interested in nonhuman organisms interactions (Connor 1995; Sachs *et al. 2004*; Bshary and Bronstein 2005, Noe and Hammerstein 1994, 1995).

In the last twenty years, as underlined by Noe (2006) in his quite recent review on the theme, many studies have been run on cooperation observing two conspecifics that could get a reward through cooperation, mainly addressed by "three different motivations: (1) detecting the mechanistic basis of naturally occurring forms of cooperation; (2) analyzing behavioural strategies specific to cooperation; and (3) testing game-theoretical models" (Noe, 2006).

In many cases these latter experiments study cooperation by using very abstract and artificial conditions to which animals, vertebrates in many cases, must be trained to, thus making it difficult to distinguish if dyads are "coordinated trough communication or acting apart together" (ibidem). It seems therefore quite relevant trying to understand how communication allows dyads to cooperate indeed. This issue can be approached to in a comparative natural and artificial behavioural science (Ponticorvo *et alii*, 2006) in which artificial organisms, such as robots or simulated agents are compared with natural organisms. In our approach we use Artificial Intelligence and Robotics tools to build artificial systems that are able to simulate, at least for some aspects, animal or human behaviour. This methodology allows us to deal with theoretical goals, because the reconstruction, both in simulation or with physical artifacts, of a model about a certain phenomenon allows to verify synthetically its explicatory potential. In particular we use Evolutionary Robotics techniques(Nolfi and Floreano, 2000), allowing the robot-environment system to self-organize and then analyze how it came to a solution.

This methodology, in recent years, has been widely applied to the emergence of communication(for a review see: Cangelosi and Parisi, 2002; Kirby, 2002; Steels, 2003; Wagner *et al.*, 2003; Nolfi, 2005). In this paper we follow the theoretical and methodological approach used by Di Paolo (1997, 2000), Quinn (2001), Quinn *et al.* (2001), Baldassarre *et al.* (2002), Trianni and Dorigo (2006), Marocco and Nolfi (2007) with one main differences: these models are mainly on quite abstract and ideal phenomena and tasks about the emergence of communication and cooperation. On the contrary our aim is to establish a strong link with phenomena and tasks derived from experiments on animal behaviour in order to get insight from this kind of data reciprocally. For this reason we model a well-defined experimental set-ups, that has been widely used in animal behaviour literature and try to compare what happens in corvids' cooperation with what happens in robots'cooperation.

In the present study we propose a model that replicates in the main aspects the 'loose string' paradigm derived from the most popular paradigm used in game-theoretical model: the Prisoner's Dilemma, applied to comparative research (Ashlock *et alii*, 1996; Clements and Stephens, 1995; Gardner *et alii*, 1984). In the 'loose string' task two agents, for example two rooks (*Corvus frugilegus*), must cooperate to obtain a reward, i.e. food, which is clearly visible, but not directly reachable. The dyad gets the reward if the two tips of a string are pulled at the same time. In the present study we model this task with artificial organisms to verify the emergence and maintenance of cooperation in artificial organisms.

2. Method

2.1. *The loose string task*

In the loose string task two members of a dyad are trained to pull a string to reach a reward. In a first phase, the agents, for example, corvids such as rooks (Scheid and Noe, in preparation), are trained separately to pull the string which allows the bird the get the food by itself. In the cooperation testing phase, the two birds could get the reward only if they pulled the string at the same time.

In this comparative study the researchers observed if the one bird pulled the string alone or two birds pulled it together, thus successfully accomplishing the task. They recorded the delay before each bird reach the apparatus where there was the string. They recorded if a bird reached the string before the other and then the other bird joined the first. Then they recorded whether one bird pulled the string alone or if the two birds pulled it at the same time and if they successfully accomplished the task. They also recorded how many pieces of food each bird did eat.

We reproduced this natural experimental task with simulated robots.

2.2. *Experimental setup*

The experimental setup involves two robots situated in an arena consisting of a square room in which robots begin each trial and of a wide corridor with a target area in it, marked by a landmark in its center. Once the robots have reached the target area, the first landmark is removed and three targets are placed in the corridor. To accomplish the task robots have to drive towards the same target. This task is derived by the "loose string" task described above and represents a situation in which the robots should coordinate themselves/cooperate to get a reward. In particular the first phase (when the robots reach the target area) represents the apparatus approaching while the second phase (when the robots have to approach the same target) represent the cooperative task in the rooks' experiment.

2.3. *The environment and the robots*

The environment consists of a 150x150 cm square arena joint with a 90x90 cm square arena both surrounded by walls (Fig.1). The corridor presents a target area and three landmarks whose positions are randomly assigned when the robots enter in the area. The robots are two e-Puck robots (Mondada and Bonani, 2007, see Fig.2) with a diameter of 7.5 cm provided with 2 motors which control the 2 corresponding wheels, 8 infrared proximity sensors located around the robot's body, a VGA camera with a view field of 240 degrees wide and 30 cm long pointing in the direction of forward motion and a LED ring on the circumference. The camera and the LED ring can be used to send and receive signals.

The environment consists of a 100x80 cm rectangular arena joint with a 60x100 cm corridor both surrounded by walls (Fig.1). The corridor presents in its second half a target area marked by a landmark in the center. When both robots enter the target area the first landmark is removed and three targets are placed at a fixed distance from the target area, in random locations but with constant distance between each other. Landmarks consist of 2cm radius coloured cylinders. The robots are two e-Puck robots (Mondada and Bonani, 2007, see Fig.2) with a diameter of 7.5 cm provided with 2 motors which control the 2 corresponding wheels, 8 infrared proximity sensors located around the robot's body, a VGA camera with a view field of 36 degrees pointing in the direction of forward motion and a LED ring on the circumference. The camera and the LED ring can be used to send and receive signals.

2.4. *The neural controller*

The neural controller of each robot is provided with sensory neurons, internal neurons with recurrent connections, motor neurons and some neurons controlling leds. These neurons allow to receive and produce signals that can be perceived by another robot. In detail in the sensory layer there are neurons that encode activation of infrared sensors and camera units; in the hidden layers there are 8 hidden neurons with recurrent connections; in the output layer there are two units that control wheels and 8 units each controlling the state of a single led, with the exception of the 4th of them controlling the hardware coupled rear leds. In Fig. 3 we present schematically the neural architecture we used in our simulation.

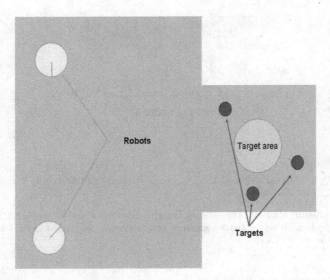

Fig. 1. The environment. There is a big rectangular arena where the robots start their trials. When they are on the target area in the corridor three targets appear and robots have to reach the same target to get a reward.

Fig. 2. The e-Puck robot with 2 motors, 8 infrared sensors, a camera and the LED ring.

2.5. *The evolutionary algorithm*

An evolutionary technique is used to set the weights of the robots' neural controller. The initial population consists of 100 randomly generated genotypes that encode the connection weights of 100 corresponding neural networks. Each genotype is translated into 2 identical neural controllers

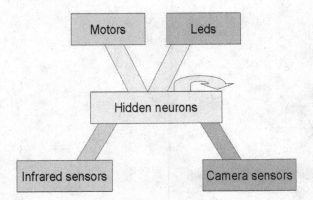

Fig. 3. The neural network: the sensory layer (activation of infrared sensors and camera units), the hidden layers (with recurrent connections); the output layer (controlling wheels and the state of leds.

which are embodied in 2 corresponding robots situated in the environment (i.e. teams are homogeneous). The 20 best genotypes of each generation are allowed to reproduce by generating 5 copies each, with 2% of their bits replaced with a new randomly selected value. The evolutionary process lasts many generations (i.e. the process of testing, selecting and reproducing robots is iterated 1000 times). The experiment is replicated 10 times each consisting of 4 trials (2000 cycles each) with 4 different starting positions in the corners of the rectangular arena . We used the following fitness function, in pseudo-code, to evolve robots:

If (a robot is on target area) Fitness +=0,25;

If (two robots are on target area)

{

Central landmark disappears and three targets appear;

If (two robots are close to the same landmark and therefore close to each other)

Fitness +=2;

}

3. Results and Conclusions

Results show that cooperation between robots is regulated by interaction between robots, with communication as a medium. In our simulative scenario the emergence of communication leads to a coordinated cooperation behavior that is somewhat similar to cooperation observed in natural organisms as corvids.

We considered some behavioural indexes that were quite similar to the ones used in the work with corvids (Scheid and Noe, in prep.). As already mentioned before, in accordance with the animal study we observed if the one robot reached the target alone or two robots reached it at the same time, thus making the targets appear. We also considered the delay before each robot reach the target area and if a robot reached the target area before and then the other followed the first. We also observed if the two robots reached the same target together solving efficiently the task.

The 10 evolved dyads proved to be able to accomplish the task, showing an efficient behaviour both in the first phase when they have to reach the target area and in the second phase when they have to approach one of the three targets together.

Let's now analyze the prototypical strategy of the best performing dyad, e.g. that gets one of the highest fitness score (in rooks' term that gets more food). This good-performing dyad shows the following behaviors. In the first phase, both robots reach the target area in the corridor that is clearly distinguished because of the landmark. In this phase leds are off. As soon as they both reach this target arena (it takes approximately 300 cycles) they turn led on. This works as a signal as they can now see each other (Fig.4a,4b,4c).

At this point the three targets appear: if the robot did not cooperate they could go to different targets. But this doesn't happen as they use this strategy: one of the robot moves to one of the target and the other follows it where closely, so they both reach the same target solving efficiently the task. This movement is leading-following: one robot moves toward the target and the other follows it. When they are close to the target, where they receive the reward, they stay close to it and close to each other. The robots also show an obstacle-avoidance behaviour that allows robots not to bump into walls and into each other and is produced when the frontal infrared sensors of the robot are activated.

In this dyad, thus, the emergence of communicative signals, turning on and perceiving leds' activation that allow the exchange between robots, leads to the emergence of this simple form of cooperation.

It's worth noting that in the present study we analyze dyads of simulated agents which are embodied and situated and that autonomously develop this simple kind of communication while they interact with a simulated physical environment. This attempt to study the evolution of communica-

186

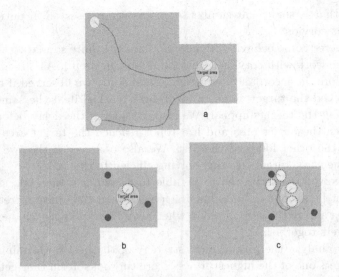

Fig. 4. The behavioural strategy. The robots reach the target area with led off (Fig. 4a), then they turn led on (dotted edge) and are visible with the camera (Fig. 4b). At this point one robot goes toward a target and the other follows it (Fig. 4c).

tion and cooperation through computational and robotic models has the important advantage to study how communication signals emerge from the grounding in robots'sensory-motor system and how cooperation can evolve and adapt to variations of the environment made up by physical environment and the other robot in the dyad.

In this paper, we use Artificial Life systems as labs in which to test hypothesis to understand a general function, in this case cooperation comparing different living forms, biological organisms with artificial organisms. These preliminary results require a wider investigation in order to understand the artificial neural mechanisms underlying these behavioural modules that result in a cooperative behaviour.

Acknowledgements

This work was supported by Cooperation in Corvids (COCOR), Collaborative Research Project (CRP) in the framework of the ESF-EUROCORES programme TECT (The Evolution of Cooperation and Trading).

References

1. Emery, N.J. (2004) Are corvids 'feathered apes'? Cognitive evolution in crows, jays, rooks and jackdaws. In: Watanabe S (ed) Comparative analysis of minds. Keio University Press, Tokyo, pp 181-213
2. Emery, N.J., Clayton, N.S. (2004a) Comparing the complex cognition of Birds and Primates. In: Rogers LJ, Kaplan G (eds) Comparative vertebrate cognition: are primates superior to non-primates? Kluwer Academic/Plenum, New York, pp 3-46
3. Emery, N.J., Clayton, N.S. (2004b) The mentality of crows: convergent evolution of intelligence in corvid and apes. Science 306:1903- 1907
4. Zucca, P., Milos, N., Vallortigara, G. (2007) Piagetian object permanence and its development in Eurasian Jays (Garrulus glandarius). Animal Cognition, 10:243-58.
5. Bugnyar, T. (2007) An integrative approach to the study of ToM-like abilities in ravens. Japanese Journal of Animal Psychology, 57, 15-27
6. Scheid, C., Noe, R. (in prep.) Implications of individual temperament in a cooperative task
7. Connor, R. C. (1995). Altruism among non-relatives: alternatives to the Prisoners Dilemma. Trends in Ecology and Evolution, 10, 8486.
8. Sachs, J. L., Mueller, U. G., Wilcox, T. P., Bull, J. J. (2004). The evolution of cooperation. Quarterly Review of Biology, 79, 135160.
9. Bshary, R., Bronstein, J. L. (2005). Game structures in mutualistic interactions: what can the evidence tell us about the kind of models we need? Advances in the Study of Animal Behavior, 34, 59101.
10. Noe, R., Hammerstein, P. (1994). Biological markets: supply and demand determine the effect of partner choice in cooperation, mutualism and mating. Behavioral Ecology and Sociobiology, 35, 111.
11. Noe, R., Hammerstein, P. (1995). Biological markets. Trends in Ecology and Evolution, 10, 336339.
12. Noe, R. (2006). Cooperation experiments: coordination through communication versus acting apart together. Animal Behaviour, 71, 1-18
13. Ponticorvo, M., Walker, R., Miglino, O. (2006). Evolutionary Robotics as a tool to investigate spatial cognition in artificial and natural systems, . In Loula A.C., Gudwin R., Queirz J., Artificial Cognition Systems, Idea Group. (p.). Hershley, PA
14. Nolfi, S., Floreano D. (2000). Evolutionary Robotics: The Biology, Intelligence, and Technology of Self-Organizing Machines. Cambridge, MA: MIT Press/Bradford Books
15. Cangelosi, A., Parisi, D. (Eds.) (2002). Simulating the Evolution of Language. London: Springer-Verlag
16. Kirby, S. (2002). Natural Language from Artificial Life. Artificial Life, 8(2):185-215.
17. Steels, L. (2003) Evolving grounded communication for robots. Trends in Cognitive Science. 7(7): 308-312.

18. Wagner, K., Reggia, J.A., Uriagereka, J., Wilkinson, G.S. (2003). Progress in the simulation of emergent communication and language. Adaptive Behavior, 11(1):37-69.

19. Nolfi, S. (2005). Emergence of Communication in Embodied Agents: Co-Adapting Communicative and Non-Communicative Behaviours. Connection Science, 17 (3-4): 231-248.

20. Di Paolo, E.A. (1997). An investigation into the evolution of communication, Adaptive Behavior 6 (2): 285-324.

21. Di Paolo, E.A. (2000). Behavioral coordination, structural congruence and entrainment in a simulation of acoustically coupled agents. Adaptive Behavior 8:(1): 25-46.

22. Quinn, M. (2001). Evolving communication without dedicated communication channels. In Kelemen, J. and Sosik, P. (Eds.) Advances in Artificial Life: Sixth European Conference on Artificial Life (ECAL 2001). Berlin: Springer Verlag.

23. Quinn, M., Smith, L., Mayley, G., Husbands, P. (2003) Evolving controllers for a homogeneous system of physical robots: Structured cooperation with minimal sensors. Philosophical Transactions of the Royal Society of London, Series A: Mathematical, Physical and Engineering Sciences 361:2321-2344.

24. Baldassarre G., Nolfi S., Parisi D. (2003). Evolving mobile robots able to display collective behaviour. Artificial Life, 9: 255-267.

25. Trianni V., Dorigo M. (2006). Self-organisation and communication in groups of simulated and physical robots. Biological Cybernetics, volume 95, pages 213-231, 2006.

26. Marocco D., Nolfi S. (2007). Emergence of communication in embodied agents evolved for the ability to solve a collective navigation problem. Connection Science, 19 (1): 53-74.

27. Ashlock, D., Smucker, M. D., Stanley, E. A., Tesfatsion, L. 1996. Preferential partner selection in an evolutionary study of Prisoner's Dilemma. Biosystems, 37, 99-125.

28. Clements, K. C., Stephens, D. W. 1995. Testing models of nonkin cooperation: mutualism and the Prisoner's Dilemma. Animal Behaviour, 50, 527-535.

29. Gardner, R. M., Corbin, T. L., Beltramo, J. S., Nickell, G. S. 1984. The Prisoner's Dilemma game and cooperation in the rat. Psychological Reports, 55, 687-696.

30. Mondada F., Bonani M. (2007). The e-puck education robot. http://www.e-puck.org/

CARWIN42: EVOLUTION OF ARTIFICIAL INTELLIGENCE CONTROLLER AND AEROMECHANICAL SETUP IN SIMULATED RACE CARS

ING. PAOLO PINTO, PHD

Dipartimento di Progettazione e Gestione Industriale, Università Federico II di Napoli;
(paolo.pinto@gmail.com)

ING. MAURO DELLA PENNA

Department of Control and Simulation, Faculty of Aerospace Engineering, Delft
University of Technology (maurodp84@hotmail.com)

ING. EMANUELA GENUA

Department of Aerodynamics, Faculty of Aerospace Engineering,
Delft University of Technology (emanuelage@msn.com)

ING. MAURIZIO MERCURIO

Dipartimento d Informatica, Sistemistica e Comunicazione, Politecnico di Milano
(dott.mercurio.maurizio@gmail.com)

ING. PASQUALE MEMMOLO

Dipartimento di Elettronica e Telecomunicazioni, Università Federico II di Napoli
(pasquale.memmolo@tin.it)

A racecar simulation software allowing evolution of driver AI and aerodynamic and mechanical setup parameters is presented. The fundamental requisites for this kind of AI are outlined, as well as several implementations used in the simulator. Major challenges posed by race cars peculiarity, such as instabilities caused by ground effect are described. The paper also shows preliminary results in the investigation of the relative importance of AI and setup towards laptime performance.

1. Introduction

Race cars change almost each time they are used. Suspension, gears, aerodynamics, are continuously played around with, in search of a track dependant optimum that also varies with the driver's capabilities and predilections. The present work describes some preliminary results obtained by CARWIN42, a software developed by a team led by Dr. Pinto and based (during the two years of its initial development phase) at the Industrial Design and Management Department of University Federico II, Napoli. CARWIN42 includes an Eulerian simulator of racecar physics, an AI engine to drive the car and an evolution engine which acts both on AI parameters and on the car's aeromechanical setup in order to reduce laptimes. While some papers about racecar setup optimization were published [7,8], to the Authors' knowledge this may be the first attempt to study its interactions with a driving AI.

2. Racecar physics

Racecar physics are dominated by extremely non-linear phenomena, and are also supposed to operate usually in regions where nonlinearities are most evident

2.1. *Tyres*

Tyres generate a force in the ground plane, the magnitude of which varies with vertical load, slip angle (angular difference between the direction of tyre speed and the tyre's equatorial plane) and slip coefficient (measure of difference between tyre peripheral speed and road speed) . Variation with these factors is non linear; most notably, increasing the vertical load gives a less than linear increment in forces. Strong negative correlation exists between longitudinal and lateral component. A model to calculate tyre forces is Pacejka's "Magic Formula" [2].

2.2. *Aerodynamic setup*

Racecars are usually designed to provide downforce, in order to augment the vertical load on tyres, thus increasing the forces they are able to produce and improving cornering, braking and low-speed acceleration. Front-rear downforce distribution heavily influences the car's behavior, the more as speed increases.

Wings generate downforce at the price of an increment in aerodynamic drag. A given wing will produce more downforce (at a diminishing rate) and drag (at an increasing rate) as its angle of attack is increased.

Modern racecars derive about 40% of their downforce from ground effect. Ground effect is more efficient than wings; that is, a given quantity of downforce will cost less in terms of drag. This downforce is obtained by shaping the car bottom in such a way that the air under the car is accelerated, therefore reducing its pressure. The magnitude of this force increases as ground clearance decreases (a role is also played by the car's pitch angle). At a critical distance from the ground, anyway, it will drop.

Static ride height selection should be such that ground clearance is minimized yet the critical value is never reached despite suspension movements. The application point of ground effect downforce is affected by the car's pitch angle too. This has unpleasant consequences in some circumstances. Under braking, infact, the car's nose gets nearer to the ground, while the rear lifts; this produces a sudden advance of the center of pressure, and a tendency for the car to spin [5].

2.3. *Mechanical setup*

Mechanical setup includes suspension tuning, front-rear brake force distribution and gear ratios selection.

A critical factor in suspension tuning is roll stiffness distribution. When a car negotiates a bend, it experiences an overturning moment that tends to load the outside tyres more than the inside tyres. The front-rear distribution of this moment is dependant on the roll stiffness of each end. Due to the nonlinear relationship between vertical load and lateral force generated by tyres, the end with greater roll stiffness will experience a greater decrease in adherence than the other. This will influence the behaviour of the car, influencing it towards oversteer, if rear end has less adherence, or understeer in the opposite case. A similar pehenomenon, longitudinal load transfer, causes, with most suspension geometries, the car to dive under braking and squat under acceleration.

A racecar suspension is designed to tackle two tasks: provide a reaction to road irregularities that is small enough to minimize adherence loss caused by varying loads, and provide a front-rear load transfer that will maximize cornering capabilities. Springs also control ground clearance variations induced by load transfer and downforce (see 2.4).

While several formulas exist to calculate an "ideal" gear ratios distribution for production cars, nothing of the sort applies to racecars. This happens because their more powerful engines easily cause low speed wheelspin, rendering too short gears useless, and because the optimum balance between top speed and acceleration varies with aerodynamic drag and also from a racetrack to another.

Harder braking will shift more load on the front wheels and require more forward-biased brake distribution. Yet on twisty tracks it can be helpful to have more braking on the rear to help the car turn in.

2.4. *Aeromechanical interactions*

In order to minimize height variations, ground effect cars require stiffer springs than normal. Alas, these cause a sizeable reduction (10% and more) of the time averaged adherence coefficient in a given condition.

3. Racecar driving

A race driver will always attempt to extract maximum forces from tyres, balancing their lateral and longitudinal component. Steering is required to keep the car as near as possible to optimal trajectory. An optimal trajectory for a single corner is, in first approximation, an arc that starts on the track side corresponding to the external of the corner, touches apex at mid corner and then goes wide again at the corner exit. This path will have a radius much greater than the corner's, allowing the car to travel faster through it.

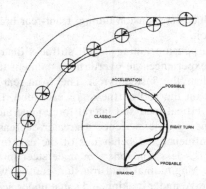

Figure 1. A racecar's trajectory. The car enters from top-right , using all the tyre's force to steer, and progressively accelerates until exit, bottom left (image taken from [1])

Very often, anyway, optimal trajectories end up not being circular arcs traveled at constant speed, but more complex curves the first part of which is traveled under braking while in the last part the car is accelerated. Drivers tend to favor understeering setups on faster tracks.

4. Software implementation

CARWIN42 has been developed in Fenix language (http://fenix.divsite.net). Fenix is an open source freeware programming language with elements from C and Pascal; it has very good and easy to use graphic libraries which were useful to reduce development time, and outweighted the language's greatest shortcoming i.e. the lack of a double precision data type.

5. Physical models

CARWIN42 can employ three different physical models for vehicle dynamics, of different accuracy and proportionally high computational effort. All models share the same aerodynamics calculation module, which includes ground effect.

Single track model is based on the assumption that both wheels have the same slip angle [3,6]. Tyre forces on the same axle are added together in "axle characteristics". These are usually calculated by point interpolation. In CARWIN42 it was decided to approximate them by Pacejka's Magic Formula. The coefficients are found by Genetic Algorithms. The use of this vertical load sensitive formula allows fast calculation of downforce effects [4].

A more accurate model computes forces for each tyre, taking into account their distance from the car's centerline. Springs are modeled in a quasi static way; that is, they instantly reach an equilibrium position under a given load. Tyre adherence coefficient is decreased as spring stiffnesses increase This model was used in the test runs shown in the present work.

The computationally heaviest and most accurate model includes dynamic behavior of springs, stabilizer bars and dampers, and transient tyre response.

6. Driver AI

Several AI controllers have been developed for CARWIN42 in the course of time. Albeit more advanced ones are now implemented, we will describe Controller Mk3, which was used in the experiments here presented.

Mk3 would only use information collected in 5 points inside a 30° wide cone ahead of it or anyway in the car's immediate surroundings (Fig. 32). This information should be only of the kind "this pixel is track – this pixel is grass" and it would be used to decide how much to steer, brake or accelerate.

Mk3 uses a fixed gearchange algorithm, that commands an higher gear if engine rpm goes above maximum power rpm, and a lower gear if it goes below maximum torque rpm.

The car is trailed by 4 sensors which are constantly kept at the track-grass interface. These determine (Fig.2) distance d from grass ahead of the car and an approximate calculation of next corner radius R , as well as its direction.

When the quantity $\dfrac{v^2 - k\sqrt{R} \cdot v}{d}$, where v is speed and k is an evolvable constant, falls below or exceeds a given value, acceleration or braking are commanded . Acceleration –braking will be limited if wheelspin exceeds given thresholds.

The desired steering angle δ is determined by the car's distances a,b, from the track sides and their increments over last calculation steps , Δa, Δb, as well as by evolvable constants p, c.

$$\delta = (b \cdot (p-1) - a \cdot (p+1)) + (\Delta a - \Delta b) \cdot c$$

The steering angle change rate is also an evolvable parameter.

Mk3 made the car quite good at preparing turns, and at negotiating chicanes, but it was still sub-optimal in corner exit. Also, it allowed for soft braking in a straight line, in most situations an anathema to racing.

The corner radius calculation system proved to give inconsistent readings, causing sudden changes in car speed. Moreover, the system was incapable of discerning between two corners having the same radius but different lengths, and therefore different optimal trajectories.

Another shortcoming was inability to take into account the increasing adherence caused by downforce as speed increases. The resulting driving' was thus' overconservative at high speeds.

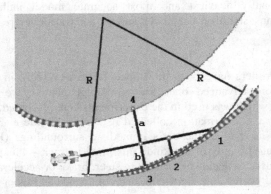

Figure 2. Sensors on AI Mk3. Sensor 1 determines distance to grass ahead of the car; sensors 3,4 determine distance a,b from the track's sides; the circumference passing through sensors 1,2,3 provides an estimate R of corner radius.

Interestingly, the software showed constant evolution of oversteering setups when this AI was used. We subsequently discovered that, this way, the car could "see" a longer free distance, thus allowing the controlooer to command faster speeds in corners. The phenomena was more evident in faster corners, where downforce effect, the beneficial influence of which was neglected by controller design, became more important.

7. Evolution engine

CARWIN42 starts by generating a car population with random parameters starting from the same position. Parameters are encoded in two linear chromosomes, one for setup and one for AI controller. After all cars have completed a lap or went off track, their genomes are evaluated. Fitness is a diminishing function of laptime; cars that did not complete a lap get a fitness proportional to the distance traveled. The slowest of the cars that completed a lap will always have a fitness greater than the best one of non-finishers.

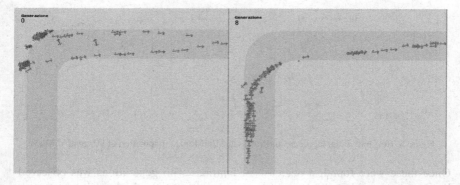

Figure 3. Learning to drive. At generation 0 cars cannot make it through the first corner. By generation 8 most of the population is selecting a racing trajectory of sort (AI Mk4, 101 cars)

Tournament selection, double point crossover, Gaussian crossover and mutation are then used to obtain a new generation. Best individual is retained.

Gear ratios are evolved in a way that made sure that no impossible configurations (i.e. these having less teeth in inferior gears) were used. In order to obtain this, it has been deemed convenient to evolve the number of teeth in highest gear and the increase in this number from each gear to the next.

8. Tests and results

Preliminary tests were carried on 3 different tracks: an High Speed Oval (HSO), a Low Speed Oval (LSO) and an half scale version of Monza track called MiniMonza (MM). We show here some examples of the results.

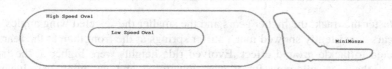

Figure 4. Circuits (same scale): High Speed oval, Low Speed Oval, MiniMonza

8.1. *Decreasing laptimes*

All tests showed a constant decrease in laptimes with generations. On the most twisty and difficult track, Minimonza, usually it took a while to generate at least one individual able to complete a lap; the selection therefore initially rewarded the individuals able to complete the most meters before leaving the track.

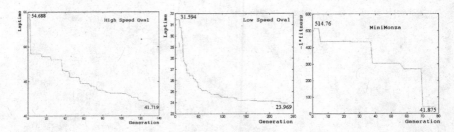

Figure 5. Best individual laptimes (antifitness at MiniMonza), population of 90 cars, AI Mk3

Once these were found, a decrease of laptimes with generations was observed here too, yet with a less gradual decrease due to the higher chance for an innovative individual to crash.

8.2. *Realistic setups*

CARWIN42 evolved setups consistent with theoretical requirements:

Table 1. Evolved setups for three different tracks

Setup Parameter	HSO	LSO	MM	Setup Parameter	HSO	LSO	MM
Front Wing Angle	5°	9°	20°	% rear braking	29%	19%	44%
Rear Wing Angle	3°	14°	12°	1st gear teeth	51	53	65
Front Springs (KN/mm)	2.19	1.39	2.48	2nd gear teeth	40	45	52
Rear Springs (KN/mm)	0.56	0.61	1.08	3rd gear teeth	33	36	42
Front Ride Height (mm)	25	21	27	4th gear teeth	24	27	33
Rear Ride Height (mm)	15	12	19	5th gear teeth	19	20	27

The faster the track, the higher gears and the smaller the selected wing angles. The cars consistently showed much stiffer springs at the front than at the rear, as expected to tackle ground effect. Evolved ride heights were higher at the front than at the rear, allowing for higher downforce and lessening aerodynamic instability phenomena.

It is also evident, though, that the AI controller behavior dictated some setup choices not in line with what would have happened with a proper driver. While it would have been expected to see an increase in rear wing angle versus front wing angle on faster circuits, this did not happen. Actually, an oversteering setup was developed on HSO to compensate for controller shortcomings (see 6). Also, front-rear braking distribution does not appear to follow a discernible path.

8.3. *Is it in the car or in the driver?*

It was also attempted to find a measure to understand if decreasing laptimes depended more on setup or AI evolution. This was done by applying a formula that measured an average variation V of parameters from lap to lap:

$$\frac{1}{N_J} \sum_{i=1}^{N_J} \frac{|P_i(G) - P_i(G-1)|}{\Delta P_i}$$

Where N is the number of cars, $P_i(G)$ is the i-th parameter's value at lap G, and ΔP_i the difference between the allowed maximum and minimum of that parameter.

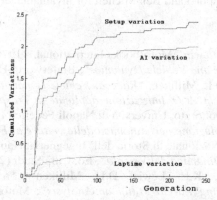

Figure 6. Cumulated function of relative variations (compared to the lap before) of setup, AI and laptime (example taken from Low Speed Oval tests, AI Mk3)

It was found out, as shown in Fig.6, that a given decrease in laptime always involved smaller relative variations of AI controller parameters than of setup parameters. This was interpreted as a greater relative influence of AI controller parameters.

9. Future work

Work is ongoing on a series of tests (enough to draw statistic inferences) using a wider variety of racetracks and new Mk4 AI controller, the which is based on a Finite State Machine and is completely different from (and far more efficient than) its predecessors.

One of the main problems encountered by our AI controllers is robustness. Even if good parameters are found, changing the starting position by just a few meters would more often than not produce an accident. Therefore subsequent work will be aimed at robustness, trying to evolve cars and AIs insensitive to

198

starting conditions. This might be achieved by putting more copies of the same car in each generation, having each start in a slightly different position. Competition between different AI concepts will also be tested.

Acknowledgments

The Authors would like to acknowledge the assistance provided by a number of Professors from Federico II University, Naples, namely Prof. Antonio De Iorio and Prof. Esamuele Santoro, Department of Industrial Design and Management and Prof. Carlo Sansone, Department of Computer Sciences and Systems. A thanks goes also to Prof. Carlo De Nicola, Department of Aeronautical Design, for the continued encouragement. Our gratitude also goes to Eng. Ricardo Divila (Nissan Motorsport) and Ben Michell for invaluable technical tips.

References

1. C. Smith, *"Tune to win"* Motorbooks International, (1978).
2. H. Pacejka, *"Tire and Vehicle Dynamics"*, Butterworth-Heinemann (2005).
3. W.F. Milliken, D.L. Milliken, *"Racecar Vehicle Dynamics"*, SAE (1994).
4. P. Pinto, *"Tecniche di Ottimizzazione applicate ai Trasporti"*, PhD Thesis, XVIII Ciclo di Dottorato, Università di Napoli Federico II, (2005).
5. P. Pinto, *"L' evoluzione aerodinamica delle vetture di Formula Uno"*, Atti del I Congresso Nazionale di Storia dell' Ingegneria, Napoli, (2006)
6. M. Guiggiani, *"Dinamica del Veicolo"*, Ed. Città Studi (2006)
7. E.M. Kasprzak, K.E. Lewis, D.L. Milliken *"Steady-state Vehicle Optimization Using Pareto Minimum Analysis"*, Motorsport Engineering Conference Proceedings, Vol1, P340-1 (2006)
8. C.D. McAllister, T.W. Simpson, K. Hacker, K. Lewis, *"Application of multidisciplinary design optimization to racecar analysis"* AIAA 2002-5608, 9th AIAA-ISSMO Symposium, Atlanta, Georgia (2002)
9. E. Genua, M. Della Penna *"Sviluppo di un simulatore di vetture da competizione dotato di pilota artificiale ed ottimizzazione genetica dei parametri di messa a punto e di guida"* Graduation Thesis in Aerospace Engineering, Tutor E. Santoro, Co-Tutor. P. Pinto, Università Federico II di Napoli, AA 2005/2006

PART V

SOCIAL SCIENCES

DISTRIBUTED PROCESSES IN A AGENT-BASED MODEL OF INNOVATION

LUCA ANSALONI, MARCO VILLANI, DAVID LANE

Department of Cognitive and Quantitative Sciences,
University of Modena and Reggio Emilia,
viale allegri 9 Reggio Emilia, I-42100, Italy
email:{luca.ansaloni, marco.villani, david.lave}@unimore.it

In this work we investigate the conditions influencing the creation of novelties and their diffusion through networks composed by agents interacting via the exchange of artifacts. By means of simulation we verified that the presence of stereotyped routines deeply influences (negatively) the robustness properties of the system, whereas the impact of strong spatial limitations or of a particular kind of direct information exchange (a request system) have more complex consequences, not all aligned to the same direction. None of these results is obvious, nor can it be simply deduced from the qualitative theory. Therefore, the simulations could make possible comparisons between the model behaviors and the theory claims, indicating new ways of improvement and development.

1. Introduction

During last decades, innovation has become a hot topic in a variety of social contexts, including technology, commerce, social systems, economic development, and policy construction. There are therefore a wide range of approaches to conceptualizing innovation in the literature.[1] A consistent theme may be identified: innovation is typically understood as the successful introduction of something new and useful, for example introducing new methods, techniques, or practices or new or altered products and services. Modeling such processes is a difficult challenge. Some authors choose the simplifying assumption that only artifacts are important (technological trajectories) whereas others claim that human creativity is the key[2-4] . But in general both agents (whatever their nature) and artifacts could be important,[5] the agents typically being the main sources of novelties and the artifacts assuring and shaping their diffusion.

There are not well established analytical tools able to capture all the features of the real situations, and more and more researchers are approaching the problem by means of simulations, whose reference frameworks derive often from complex systems theory: agent-based models, genetic algorithms, cellular automata. All these methods share the characteristic of taking into account heterogeneity and multi-level aspects, essential components of the multifaceted phenomena linked with innovation.

In particular, agent-based models (ABMs) are well suited to bridge the gap between hypotheses concerning the microscopic behavior of individual agents and the emergence of collective phenomena in systems composed of many interacting parts (for example, the emergence of new patterns of interaction due to the introduction of new objects into the system). In our work we take advantage of these aspects, whereas we leave to the initiative of each single agent the job of creating new objects. Of course, the possibility of creating and the particular design of these new objects are influenced by the position of each single agent in the whole system. In other words, in our system there is a continuous causal interaction between the individual level and the population level.

This approach has already led us to some interesting results highlighting the importance of the relationships among agents, which can influence the information flows through the system and in such a way shape the structures emerging in agent-artifact space[6–8]. The aim of the present work is to analyze in a deeper way the effects of the interaction rules acting at the agent scale, on the system as a whole.

2. The model

The model is based on an existing qualitative theory of innovation, elaborated by Lane and Maxfield (LM theory in the following). The theory provides a basic ontology for the model, its entities and their interaction modalities.[5] The model (ISCOM Innovation Model, I_2M in the following) represents therefore a simplified universe inhabited by a subset of the entities posited by the theory, and allows us to carry out simulations, which in this context have the role that experiments play in the hard sciences, as for example physic or chemistry[9].

Let us now briefly introduce the main features of the model: we will limit ourselves here to a very concise overview, and we refer the interested reader to[6–8] for a more complete and detailed account.

The first direct claim of the theory is that agents and artifacts are both important for innovation. This notion underlies the concepts of an agent-artifact space and in particular that of reciprocality, which essentially claims that artifacts mediate interactions between agents, and vice versa, and that both agents and artifacts are necessary for a proper understanding of the behavior of market systems and of innovation. One straightforward consequence of this claim is that it excludes the possibility to project agent-artifact space onto one of the two constitutive subspaces and retain sufficient structure to understand innovation dynamics.

In the LM theory, artifacts are given meanings by the agents that interact with them, and different agents take different roles with respect to transformations involving artifacts. In this context, innovation is not just novelty, but rather a modification in the structure of agent-artifact space, which may unleash a cascade of further changes.

In I_2M agents can produce artifacts, each artifact being assembled by means of a recipe (a set of input and operators); the artifacts in turn can be used by other agents to build their own artifacts. The meaning of artifacts is just what agents do with them, while the role of agents is defined by which artifacts they produce, for whom, and with which other agents they generate new kind of artifacts or production recipe. The role of agents is also partly defined by the social networks in which they are embedded. In order to better exploit their acquaintances, the agents can give an evaluation (a vote) to each existing relationship: the higher is the vote, the higher is the probability of choosing the partner to jointly realize a new artifact or production recipe.

In fact, the agents can try to widen the set of their recipes by applying genetic operators to their own recipes, or they can try to collaborate with another agent: in the latter case, in order to build the new recipe the agents can work together to manipulate all the recipes owned by either of them.

The current version of I_2M is able to deal with quantity, as it takes explicitly into account the number of items produced by each recipe. Therefore, each agent has a stockpile where its products are put, and from which its customers can obtain them. Each agent tends to a desired level of artifacts items present within its stock: if the stock level is lower than the desired one the agent increases the production of the corresponding recipe; if the stock level is higher the agent decreases the production of the recipe. Each agent has to produce each recipe at least once, and cannot produce a recipe more than a given number of times each step. Production is assumed to be fast, i.e. it does not take multiple time steps.

If the recipes output is not used by some agents, or if one of the needed inputs is not present within the fraction of the world known by the agent, a counter is incremented; otherwise the counter is set to its default. When this counter exceeds a given threshold the corresponding recipe is discarded, because it is useless or not realizable. As far as innovation is concerned, let us remark that an agent can invent new recipes or manipulate the existing ones. In the present version of the model no new agents are generated, while agents can die because of lack of inputs or of customers.

A key point is the structure of artifact space. What is required is that the space has an algebraic structure, and that suitable constructors can be defined to build new artifacts by combining existing ones. We have adopted a numerical representation for artifacts and the use of mathematical operators, instead of e.g. binary strings, ?-calculus or predicate calculus, which at this level of description introduces only unnecessary complications. Therefore the agents are producers of numbers (indicated as names in the following), by means of the manipulation of other numbers, and the recipes are defined by a sequence of inputs and operators.

The model is asynchronous: at each time step an agent is selected for update, and it tries to produce what its recipes allow it to do. So, for each recipe, it looks for the input artifacts and, if they are present in the stocks of its suppliers, it produces the output artifact and puts it into its stock (the stocks of the supplier are of course reduced). This sequence is iterated till the desired level of production is reached.

Besides performing the usual buy-and-sell dynamics, an agent can also decide to innovate. Innovation is a two-step process: in the first step, the agent defines a goal, i.e. an artifact the agent wishes add to the list of its product, while in the second step, the agent tries to generate a recipe that produces the desired artifact. In the goal-setting phase, an agent chooses one of the known types of artifacts (which, recall, is a number M) and then either tries to imitate it (i.e. its goal is equal to M) or it modifies it by a jump (i.e. by multiplying M times a random number in a given range). It has been verified that imitation alone can lead to a sustainable production economy, in which however innovation eventually halts[6] . In the goal-realizing phase (after setting its goal), an agent tries to reach its target by combining recipes to generate a new one via genetic algorithms.

3. The experiments

We are interested in investigating the conditions capable of influencing the creation of novelties and their diffusion through the system. In the past we investigated the influence of single parameters on the system performances, or some particular scenario (that is, the combinations of more than one parameter, or the introduction of new rules). In this work we present three interesting situations, involving stereotyped routines, spatial constraints and distributed design processes. All these situations are compared with the standard one, whose rules are described on the previous paragraph; in order to collect statistics we average several runs for each scenario.

In the work belong we examine two main classes of variables: extensive variables (the number of agents, the number of artifacts, the system diversity that is, the number of different artifact types contemporarily present within the system) and intensive variables (the system diversity divided by the number of agents, the median number of recipes owned by each agent, and the median production of the set of recipes owned by each agent).

3.1. *Stereotyped routines*

In order to produce new artifact, as output, the recipes use already existing artifacts as input. At the moment at which an agent wants to use one of his recipes to produce an output, it might be the case that that the stocks of the usual provider on one of the inputs is empty: if this is the case, the agent has to choose another provider from the agents he already "knows". Typically the choice takes place within a subset of the whole list of acquaintances: in other words, as the real ones, our simulated agents tend to favor other providers already employed in the past.

In reality, very often also innovation processes are influenced by the already existing routines: when a firm tries to produce a new kind of artifact, a widespread behavior is that of starting from already running processes. In fact, it is less expensive to (slightly) modify already working procedures and machines, than build totally new tools and methods.

The I_2M translation of these facts is respectively that:

- the list of acquaintances is ordered, and the search for another provider starts from the fist occurrence of this list.
- the new recipe, able to produce the desired goal, is designed by means of a genetic algorithm (that is, by means of a recombination of already existing recipes)

We modified this situation

- allowing the agents to freely choose their providers among the whole set of acquaintances, irrespectively of their order
- allowing the agents to initialize the genetic algorithm by using the whole set of known artifacts and operators, irrespectively of their real use in the current set of recipes

The result of these changes doesnt significantly modify the general system performance in terms of artifacts diversity, but deeply influence the structure of the relationships among agents. In Fig. 1c one can see the betweenness centralization indexes of the net whose nodes are the agents, two agents being linked if they exchange artifacts. As can be seen, the median number of recipes owned by each agent dramatically changes, and the same time their inhomogeneity decreases: the difference between the median and the maximum number of recipes owned by each agent becomes smaller. As a result, each artifact constitutes the input for fewer recipes with respect to the usual situation, allowing the contemporaneous existence of more recipes. The final consequence is a more robust system, where no agents die.

In particular the modification of the genetic algorithm initialization procedure, responsible for the creation of new recipes, has an even more deeper effect: despite a relatively high inhomogeneity among agents the typical recipes production levels, usually close to the inferior part of their range, reach their maximum value. In this case, the agent-artifact structure allows the exploitation of the whole system potential.

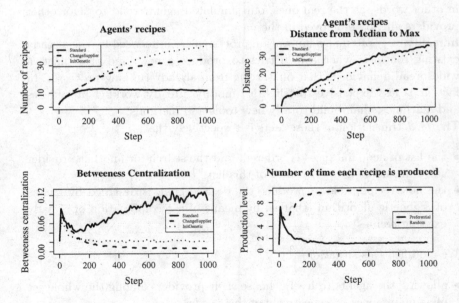

Fig. 1. Number of recipes per agent: (a) median, (b) difference between maximum and median values, (c) betweenness centralization index and (d) recipes production levels. Statistics made on 10 runs of the three main situations: normal, with random providers and with modified initialization of the genetic algorithms.

3.2. *Spatial constraints*

Untill this point in our discussion, the agents in I_2M have no relational limits: they potentially can enter into relationship with any other agents without restrictions. Each agent has its own set of acquaintances, which depends on the agents relationships (client-provider or simple acquaintance relationships) and the agents past history. This is not always the case in the real world: the environment of economic and social agents has a spatial structure, which affects agents interactions despite the presence of long-range communication media.

In order to model such spatial constraints on relationships, we cold limit the agents list of acquaintances to an *a priori* given subset. Therefore, we suppose that the agents live in a 2D environment, and that each agent can know only the agents close to him: in other word, we are building a 2D cellular automata, in which each cell is constituted by an agent having a Moore neighbourhood with a fixed radius. The opposite sides of the lattice are assumed to communicate (the so-called toroidal configuration) in order to assure the same neighbourhood size to each agent. All the already presented characteristics of I_2M hold unchanged: simply each agent acts within the defined neighbourhood (in order to maintain the acquaintanceship, spatial limitation, we block only the agents capability of sharing recipes and therefore acquaintances).

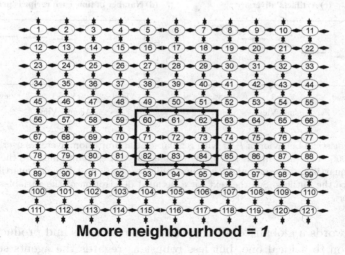

Moore neighbourhood = 1

Fig. 2.   Spatial vincula rappresentation.

We use a CA having 11x11 cells, each cell being a complete I_2M agent; the agents list of acquaintances is defined by the Moore neighbourhood of radius R, the radius being fixed for each series of experiments; the automaton stops after 10000 steps, a number of iteration sufficient to reach a stable configuration; for each radius we report statistics made on 10 runs. The classical case of global knowledge is therefore constituted by the set of experiments with R=5.

The presence of strong spatial constraints deeply influences the models behaviour. From the data shown in Fig. 3 one can see that the higher is the neighbourhood radius, the higher is the artifact diversity per agent and the average production level of the recipes, but the lower is the number of surviving agents and the number of recipes owned by each agent.

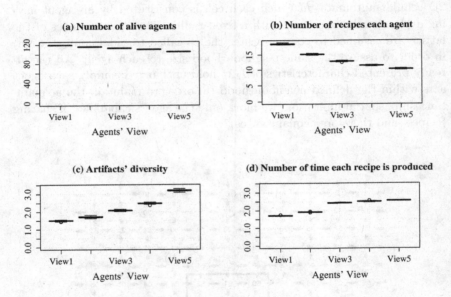

Fig. 3. Statistical behavior of I_2M with regular spatial limitation: averages over 10 runs at step 10000. The thick line is the median value, the box extremes indicate the second and third quartiles, the upper and lower line indicates the first and fourth quartiles, and the dots are the outliers. See the text for a detailed description of the variables and of its behavior.

In other words a globalised word is more heterogeneous and productive in comparison to a local one, but less robust as regards the agents survival probabilities.

It is possible to explain this apparent contradiction by observing that with low radius the system is highly fragmented, and each fragment has therefore therefore high probability of building artefacts already produced in other parts of the system. On the other hand, because of the local acquaintance structure, agents will tend to use only locally produced inputs: this prevents the loss of recipes, allowing the otherwise problematical survival of many agents. Therefore, the contemporaneous presence on global scale of many copies of the same kinds of artifact, and on local scale of few artifacts items causes the decrease of the systems diversity (see[10] for a more detailed presentation).

3.3. *Goals distributed design*

In the real world, competition between producers based on price or quality differences are in general critical in determining when a firm abandons production of a given artifact type. Neither price nor quality is directly represented in our model. For us, production from a particular recipe will cease for one of two reasons:

- the output artifact is useless (no other agent makes use of the output artifact)
- the process cant operate, because the needed input are not available

The only way the I_2M agents have to infer the usefulness of an artifact is that of creating the artifact and inserting it into the word; nevertheless, it is possible to elude these difficulties and even change them into a new opportunity. In case of absence of providers, the agents that need inputs could inform other agents about their necessities: in such a way these last agents have an additional way to evaluate the usefulness of an artifact (there are other agents that need it).

In I_2M we therefore allow the agents to publish a list of request, available to the agents with which they have a realtionship (providers and acquaintances). Each request has the same temporal duration of the corresponding recipe.

Agents that innovate and are aware of these requests have p_g probability of choosing one of the requests as their goal. In such a way, a given fraction of the innovation processes will have the requests as goal: it is an interesting distributed way of innovation design processes.

We performed several groups of experiments, each group having a given p_g value (all the agents sharing the same pg value): the aim of the experiments is to individuate the effect of p_g on the system performances. The results are not obvious, and it is possible to individuate two typical scales: very small p_g values, and higher p_g values. When p_g spans along its whole range, higher and higher p_g values correspond to lower and lower number of recipes per agent, production levels, and even agents. In effect, the fact that the agents try to build mainly just disappeared artifacts (at high p_g values) implies that the creation process of novelties is stopped, blocked in realizing only already existed objects. Therefore, always higher p_g values imply always fewer novelties, and an artifact space more and more stopped in its development.

However, at smaller scale there is another effect. When p_g is near to 0.004 we can note a small peak of the number of recipes per agent, and a definitely more marked peak of the recipes production level. That is, there is a point where the equilibrium between the need of novelties and the satisfaction of requests reaches an optimum, and can make the system more robust.

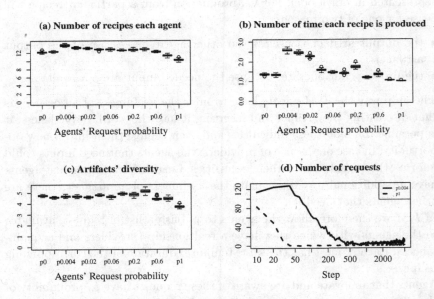

Fig. 4. Statistical behavior of I_2M with the request system active: averages over 10 runs at step 5000. (a) Number of recipes per agent, (b) recipes production (c) artifacts diversity, divided by the number of agents. The thick line is the median value, the box extremes indicate the second and third quartiles, the upper and lower line indicates the first and fourth quartiles, and the dots are the outliers. (d) The temporal behavior of the number of requests of two extreme cases (p_g=0.004 and p_g=1.0). See the text for a detailed description of the variables and of its behavior.

Last but not least, the number of requests has an interesting temporal behaviour: the higher is the probability of satisfy a request, the smaller is the period where requests are made. In fact, a request fulfilment allows the existence of a recipe that otherwise should disappear: the survival of the corresponding artifact prevents the creation of other request (caused by the disappearing of the artifact), and in few times the whole system is able to create a situation without request (by paying this favourable situation with a low initial development of the artifact space). On the contrary, an only partial request fulfilment allows a starting phase abundant of novelties, by paying this favourable situation with a much more ample period with recipes (and artifacts) disappearance (see Fig. 4).

4. Conclusions

In this work we investigate the conditions influencing the creation of novelties and their diffusion through networks composed of many agents interacting via the production and exchange of artifacts. By means of simulation we verified that the presence of stereotyped routines deeply influences (negatively) the robustness properties of the system, whereas the impact of strong spatial limitations or of a particular kind of direct information exchange (the request system) have more complex consequences, not all aligned to the same direction.

None of these results is obvious, or can be simply deduced from the qualitative theory. Therefore, the simulation allows comparisons between the model behaviours and the theory claims, indicating new ways of improvement and development. The present version of the model is only a first step toward this high objective, but the results obtained so far appear to encourage the the deepening of the already started experiments, and further development of I_2M. Future research improvements should be aimed at exploring the effect of topologies different from that of regular lattices, and simulating more complex interactions among agents and artifacts.

References

1. Fagerberg, Jan, C. David and R. R. Nelson, *Innovation: A Guide to the literature* (Fagerberg and Jan, 2004).
2. G. dosi, *Research Policy* **Technological paradigms and technological trajectories**, 147 (1982).
3. J. Schumpeter, *The Theory of Economic Development* (Harvard University Press, 1934).
4. J. G. March, *Organization Science* **Exploration and Exploitation in Organizational Learning** (1991).

5. D. Lane and R. Maxfield, *Journal of Evolutionary Economics* **Ontological uncertainty and innovation**, 3(January 2004).
6. D. Lane, R. Serra, M. Villani and L. Ansaloni, *Complexus* **A theory-based dynamical model of innovation processes**, 177 (2005).
7. R. Serra, M. Villani and D. Lane, *Complexity perspectives on innovation and social change* (Berlin, Springer, ch. Modelling Innovation.
8. M. Villani, L. Ansaloni, D. Bagatti, D. Lane and R. Serra, *Proceedings of ECCS07: European Conference on Complex Systems.* **Novelties and structural changes in a dynamical model of innovation processes** (2007).
9. A. Cangelosi and D. Parisi, *Simulating the Evolution of Language* (London, Springer, 2002), ch. Computer simulation: A new scientific approach to the study of language evolution.
10. M. Villani, R. Serra, L. Ansaloni and D. Lane, *Proceedings of the International Conference of Cellular Automata for Research and Industry (ACRI2008) Yokohama, Japan* **Global and local processes in a model of innovation** (2008).

MEET, DISCUSS AND TRUST EACH OTHER: LARGE VERSUS SMALL GROUPS

TIMOTEO CARLETTI[1]

Département de Mathématique, Facultés Universitaires Notre Dame de la Paix
Namur, B5000, Belgium
[1] *E-mail: timoteo.carletti@fundp.ac.be*

DUCCIO FANELLI[2]

Dipartimento di Energetica and CSDC, Università di Firenze, and INFN,
Firenze, 50139, Italy
[2] *E-mail: duccio.fanelli@gmail.com*

ALESSIO GUARINO[3]

Université de la Polynésie Francaise
BP 6570 Faa'a, 98702, French Polynesia
[3] *E-mail: alessio.guarino@upf.pf*

ANDREA GUAZZINI[4]

Institute for Informatics and Telematics (IIT)
Pisa, 56124,Italy
[4] *E-mail: andrea.guazzini@unifi.it*

In this paper we propose a dynamical interpretation of the sociological distinction between large and small groups of interacting individuals. In the former case individual behaviors are largely dominated by the group effect, while in the latter mutual relationships do matter. Numerical and analytical tools are combined to substantiate our claims.

Keywords: Opinion Dynamics; complex systems; social dynamics.

1. Introduction

Sociophysics is a long standing[1] research field addressing issues related to the characterization of the collective social behavior of individuals, such as culture dissemination, the spreading of linguistic conventions, and the dynamics of opinion formation.[2-6] These are all interdisciplinary applications which sit the interface of different domains. The challenge is in fact to model the dynamical evolution of an ensemble made of interacting, micro–constituents and infer the emergence of collective, macroscopic behaviors that are then eventually accessible for direct experimental inspection. Agent based computational models are widely employed in sociophysics applications and also adopted in this paper. They provide in fact a suitable setting to define local rules which govern the evolution of the microscopic constituents.

In recent years, much effort has been devoted to the investigation of social networks, emerging from interaction among humans. In the sociological literature a main distinction has been drawn between small[7] and large[8–10] groups, as depending on its intrinsic size, the system apparently exhibits distinct social behaviors. Up to a number of participants of the order of a dozen, a group is considered small. All members have a clear perception of other participants, regarded as individual entities: Information hence flows because of mutual relationships. Above this reference threshold, selected individuals see the vast majority of the group as part of a uniform mass: There is no perception of the individual personality, and only average behaviors matter. This distinction is motivated by the fact that usually humans act following certain prebuilt "schemes"[11,12] resulting from past experiences, which enables for rapid decision making whitout having to necessarily screen all available data. This innate data analysis process allows one for a dramatic saving of cognitive resources.

These conclusions have been reached on the basis of empirical and qualitative evidences.[7,13–15] We are here interested in detecting the emergence of similar phenomena in a simple model of opinion dynamics.[16,17] As we shall see in our proposed formulation agents posses a continuous opinion on a given subject and possibly modify their beliefs as a consequence of binary encounters.

The paper is organized as follows. In the next section the model is introduced. Forthcoming sections are devoted to characterizing the quantities being inspected and develop the mathematical treatment. In the final section we sum up and comment about future perspectives.

2. The model

We shall investigate the aforementioned effects related to the size of the group of interacting individuals (*social group*), within a simple opinion dynamics model, recently introduced in[16] . For the sake of completeness we hereafter recall the main ingredients characterizing the model. The interested reader can refer to[16] for a more detailed account.

We consider a closed group composed by N individuals, whose opinion on a given issue is represented by a continuous variable O_i, scanning the interval $[0, 1]$; moreover each agent i is also characterized by the so–called *affinity*, a real valued vector of $N - 1$ elements, labeled α_{ij}, which measures the quality of the relationship between i and any other actor j belonging to the community.

Agents interact via binary encounters possibly updating their opinion and relative affinity, which thus evolve in time. Once agents i and j interact, via the mechanism described below, they converge to the mean opinion value, provided their mutual affinity scores falls below a pre-assigned threshold quantified via the parameter α_c. In formulae:

$$O_i^{t+1} = O_i^t - \frac{1}{2} \Delta O_{ij}^t \, \Gamma_1\left(\alpha_{ij}^t\right) \quad \& \quad O_j^{t+1} = O_j^t - \frac{1}{2} \Delta O_{ji}^t \, \Gamma_1\left(\alpha_{ji}^t\right) , \quad (1)$$

where $\Delta O_{ij}^t = O_i^t - O_j^t$ and $\Gamma_1(x) = \frac{1}{2}\left[\tanh(\beta_1(x - \alpha_c)) + 1\right]$. The latter works as an *activating function* defining the region of trust for effective social interactions. On the other hand bearing close enough opinions on a selected topic, might induce an enhancement of the relative affinity, an effect which is here modeled as:

$$\alpha_{ij}^{t+1} = \alpha_{ij}^t + \alpha_{ij}^t(1 - \alpha_{ij}^t)\,\Gamma_2\left(\Delta O_{ij}^t\right) \quad \& \quad \alpha_{ji}^{t+1} = \alpha_{ji}^t + \alpha_{ji}^t(1 - \alpha_{ji}^t)\,\Gamma_2\left(\Delta O_{ji}^t\right) ,$$
$$(2)$$

being $\Gamma_2(x) = -\tanh(\beta_2(|x| - \Delta O_c))$. This additional *activating function* quantifies in ΔO_c the largest difference in opinion (ΔO_{ij}^t) which yields to a positive increase of the affinity amount α_{ij}^t. The parameters β_1 and β_2 are chosen large enough so that Γ_1 and Γ_2 are virtually behaving as step functions. Within this working assumption, the function Γ_1 assumes value 0 or 1, while Γ_2 is alternatively -1 or $+1$, depending on the value of their arguments [a].

The affinity variable, α_{ij}^t, schematically accounts for a large number of hidden traits (diversity,personality, attitudes, beliefs...), which are nevertheless non trivially integrated as an abstract simplified form into the model. Note also that the affinity accounts for a *memory* mechanism: indeed once two agents meet, the outcome of the interaction in part depends on their history via the affinity scores.

---

[a]We shall also emphasize that the logistic contribution entering Eq. (2) maximizes the change in affinity when $\alpha_{ij}^t \approx 0.5$, corresponding to agents i which have not yet build a definite judgment on the selected interlocutor j. Conversely, when the affinity is close to the boundaries of the allowed domain, marking a clear view on the worth of the interlocutor, the value of α_{ij}^t is more resistant to subsequent adjustments.

To complete the description of the model let us review the selection rule here implemented. Each time step a first agent i, is randomly extracted, with uniform probability. Then a second agent j is selected, which minimizes the *social metric* D_{ij}^t and time t. This is a quantity defined as:

$$D_{ij}^t = d_{ij}^t + \mathcal{N}_j(0,\sigma)\,, \tag{3}$$

where $d_{ij}^t = |\Delta O_{ij}^t|(1 - \alpha_{ij}^t)$ is the so–called *social distance* and $\mathcal{N}_j(0,\sigma)$ represents a normally distributed noise with zero mean and variance σ, that can be termed *social temperature*.[16] The rationale inspiring the mechanisms here postulated goes as follows: The natural tendency for agent i to pair with her/his closest homologous belonging to the community (higher affinity, smaller opinion distance), is perturbed by a stochastic disturbance, which is intrinsic to the social environment (degree of mixing of the population).

The model exhibits an highly non linear dependence on the involved parameters, α_c, ΔO_c and σ. In a previous work[16] the asymptotic behavior of the opinions dynamics was studied and the existence of a phase transition between a consensus state and a polarized one demonstrated. It should be remarked however that the fragmented case might be metastable; in fact if the mean separation between the adjacent opinion peaks is smaller than the opinion interaction threshold, ΔO_c, there always exists a finite, though small, probability of selecting two individuals belonging to different clusters, hence producing a gradual increase in the mutual affinities, which eventually lead to a merging of the, previously, separated clusters. This final state will be achieved on extremely long time scales, diverging with the group size: socially relevant dynamics are hence likely to correspond to the metastable regimes.

A typical run for $N = 100$ agents is reported in the main panel of Fig. 1, for a choice of the parameters which yields to a monoclustered phase. This is the setting that we shall be focusing on in the forthcoming discussion: Initial opinions are uniformly distributed in the interval $[0,1]$, while α_{ij}^0 are randomly assigned in $[0,1/2]$ with uniform distribution, parameters have been fixed to $\alpha_c = 0.5$, $\Delta O_c = 0.5$ and $\sigma = 0.01$.

Once the cluster is formed, one can define the *opinion convergence time*, T_c, i.e. time needed to aggregate all the agents to the main opinion group [b]. A second quantity T_α can be introduced, which measures the time scale for the convergence of the *mean group affinity* to its asymptotic value 1. The latter will be rigorously established in the next section.

---

[b]We assume that a group is formed, i.e. aggregated, once the largest difference between opinions of agents inside the group is smaller than a threshold, here 10^{-4}.

Such quantities are monitored as function of time and results are schematically reported in Fig. 1. As clearly depicted, the evolution occurs on sensibly different time scales, the opinion converging much faster for the set of parameters here employed.

In the remaining part of this paper, we will be concerned with analyzing the detail of this phenomenon highlighting the existence of different regimes as function of the amount of simulated individuals. More specifically, we shall argue that in small groups, the mean affinity converges faster than opinions, while the opposite holds in a large community setting. Our findings are to be benchmarked with the empirical evidences, as reported in the psychological literature. It is in fact widely recognized that the dynamics of a small group (workgroup) proceeds in a two stages fashion: First one learns about colleagues to evaluate their trustability, and only subsequently weight their input to form the basis for decision making. At variance, in large communities, only a few binary interactions are possible among selected participants within a reasonable time span. It is hence highly inefficient to wait accumulating the large number of information that would eventually enable to assess the reliability of the interlocutors. The optimal strategy in this latter case is necessarily (and solely) bound to estimating the difference in opinion, on the topic being debated.

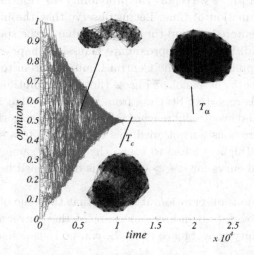

Fig. 1.   Opinions as function of time. The underlying network is displayed at different times, testifying on the natural tendency to evolve towards a coherent ensemble of affine individuals. T_c and T_α are measured according to our conventions.

3. The social network of affinities

In our model the affinity enters the selection mechanism that makes agents to interact. We can restate this fact by postulating the existence of an underlying *social network*, which drives the interactions and thus the opinion flow. In this perspective, the affinity can be seen as the *adjacency matrix* of a *weighted* [c] graph. In such a way we are formally dealing with an *adaptive* social network[18,19] : The network topology influences the opinion dynamics, the latter providing a feedback on the network itself. In other words, the evolution of the topology is inherent to the dynamics of the model because of the proposed self-consistent formulation and not imposed a priori as an additional, external ingredient, i.e. rewire and/or add/remove links according to a given probability[20,21] once the state variables have been updated. From this point of view, the mean group affinity is the *averaged outgoing degree* – called for short "the degree" in the following – of the network:

$$< k > (t) = \frac{1}{N} \sum_i k_i^t,$$ (4)

where the degree of the i–th node is $k_i^t = \sum_j \alpha_{ij}^t/(N-1)$. The normalizing factor $N-1$ allows for a direct comparison of networks made of a different number of agents. Let us observe that we chose to normalize with respect to $N-1$ because no self-interaction is allowed for.

In the left panel of Fig. 2 we report the probability distribution function for the degree, as a function of time. Let us observe that the initial distribution is correctly centered around the value $1/4$, due to the specificity of the chosen initial condition. The approximate Gaussian shape results from a straightforward application of the Central Limit Theorem to the variables $(k_i^t)_{i=1,...,N}$. In the right panel of Fig. 2 the time evolution of the mean degree $< k > (t)$ is reported. Starting from the initial value $1/4$, the mean degree increases and eventually reaches the value 1, characteristic of a complete graph. As previously mentioned this corresponds to a social network where agents are (highly) affine to each other. In the same panel we also plot the analytical curve for $< k > (t)$, as it is determined hereafter.

From the previous observation, it is clear that the time of equilibration of $< k > (t)$ provides an indirect measure of the convergence time for $\alpha_{ij}^t \to 1$. The affinity convergence time, T_α can be thus defined as:

$$T_\alpha = \min\{t > 0 :< k > (t) \geq \eta\},$$ (5)

---

[c]In fact the trustability relation is measured in terms of the "weights"$\alpha_{ij}^t \in [0,1]$.

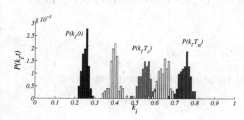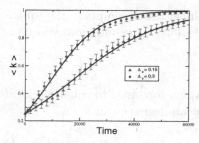

Fig. 2. Time evolution of the degree probability distribution function and the averaged degree. Left panel : Several histograms representing successive snapshots of the dynamics are displayed: $t = 0$ (black online), two generic intermediate times (wight), $t = T_c$ (red online) and $t = T_\alpha$ (blue online); histograms are normalized to unity. Right panel : $< k >$ versus time. Symbols refer to direct simulations. The solid lines are obtained as a one parameter fit of the theoretical expression Eq. (10).

where $\eta \in (0, 1)$ is a threshold quantity. The closer η is to 1 the stronger the condition on the value of α_{ij}^t (the larger the value for T_α) which identifies an affine unit. In the following we shall assume $\eta = 3/4$.

4. Results

The aim of this section is to analyze the behavior of T_α and of $< k > (t)$ so to provide an analytical interpretation for the results presented in the previous section.

From the definition of mean degree Eq. (4) one can compute the time evolution of $< k > (t)$ in a continuous time setting:

$$\frac{d < k >}{dt} = \frac{1}{N(N-1)} \sum_{i,j} \frac{d\alpha_{ij}^t}{dt} . \qquad (6)$$

To deal with the update rule for the affinity Eq. (2) we assume that we can decouple the opinions and affinity dynamics by formally replacing the activating function $\Gamma_2(\Delta O_{ij}^t)$ with a suitably tuned constant. Strictly speaking, this assumption is correct only for $\Delta O_c = 1$, in which case the opinions are always close enough so to magnify the mutual affinity scores of the interacting agents, as a results of a self consistent evolution. For the general case, $\Delta O_c \leq 1$, what we are here proposing is to replace $\Gamma_2(\Delta O_{ij}^t)$ by some effective value termed γ_{eff}, determined by the dynamics itself. The technical development that yields to the optimal estimate of γ_{eff} will be dealt in the appendix.

Under this assumption, the equation Eq. (2) for the evolution of α_{ij}^t admits the following continuous version :

$$\frac{d\alpha_{ij}^t}{dt} = \gamma_{eff}\alpha_{ij}^t \left(1 - \alpha_{ij}^t\right) , \tag{7}$$

which combined to equation Eq. (6) returns the following equation for the mean degree evolution :

$$\frac{d < k >}{dt} = \gamma_{eff} \left(< k > (t) - < (\alpha_{ij}^t)^2 >\right) . \tag{8}$$

Assuming the standard deviation of (α_{ij}^t) to be small [d] for all t, implies $< (\alpha_{ij}^t)^2 > \sim < \alpha_{ij}^t >^2 = < k >^2$, which allows us to cast the previous equation for $< k >$ in the closed form :

$$\frac{d < k >}{dt} = \gamma_{eff} \left(< k > - < k >^2\right) . \tag{9}$$

This equation can be straightforwardly solved to give:

$$< k > = \frac{k_0}{k_0 + (1 - k_0)e^{-\gamma_{eff}t}} , \tag{10}$$

where $k_0 = < k > (0)$. We can observe that such solution provides the correct asymptotic value for large t. Moreover γ_{eff} plays the role of a *characteristic* time and in turn enables us to quantify the convergence time T_α of the affinity via :

$$T_\alpha = \frac{1}{\gamma_{eff}} \log \left(\frac{\eta(1 - k_0)}{k_0(1 - \eta)}\right) = \frac{\eta'}{\gamma_{eff}} . \tag{11}$$

In the appendix we determine [e] the following relation which allows to express γ_{eff} as a function of the relevant variables and parameters of the models, i.e. T_c, ΔO_c and N:

$$\gamma_{eff} = \frac{1}{N^2} + \frac{T_c}{T_\alpha N^2}\rho , \tag{12}$$

where $\rho = -(1 + 2\Delta O_c \log(\Delta O_c) - \Delta O_c^2)$, thus recalling Eq. (11) we can finally get:

$$T_\alpha = \eta' N^2 \left(1 - \frac{T_c \rho}{N^2 \eta'}\right) . \tag{13}$$

---

[d]This assumption is supported by numerical simulations not reported here and by the analytical considerations presented in.[22]
[e]This is case a) of Eq. (A.7). In the second case the result is straightforward: $T_\alpha = \eta' N^2/(\rho + 1)$.

From previous works[16,23] we know that the dependence of T_c on N, for large N, is less than quadratic. Hence, for large N, the second term in the parenthesis drops, and we hence conclude that, the affinity convergence time grows like $T_\alpha \sim N^2$, as clearly shown in the main panel of Fig. 3. The prefactor'e estimate is also approximately correct, as discussed in the caption of Fig. 3.

The above results inspire a series of intriguing observation. First, it is implied that the larger the group size the bigger T_α with respect to T_c. On the contrary, making N smaller the gap progressively fades off. Dedicated numerical simulations (see left inset of Fig. 3) allows to indentify a turning point which is reached for small enough values of N: there the behavior is indeed opposite, and, interestingly, $T_c > T_\alpha$. The transition here reproduced could relate to the intrinsic peculiarities of the so called "small group dynamics"to which we made reference in the introductory section[7-10] . Furthermore, it should be stressed that the critical group size determining the switching between the two regimes here identified, is quantified in $N \simeq 20$, a value which is surprisingly closed to the one being quoted in social studies.

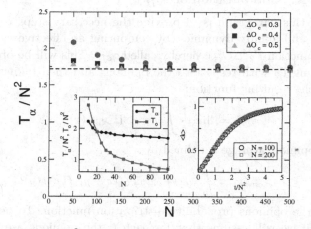

Fig. 3. Main panel: $T\alpha/N^2$ vs N for different values of the parameter ΔO_c. The data approach a constant value ($T_\alpha/N^2 \sim 1.72$) clearly indicating that the time of convergence of the affinity matrix scales quadratically with the number of agents, in agreement with the theory. The asymptotic value estimated by our theory is 2.19, the discrepancy being therefore quantified in about 15 %. Left inset: T_α/N^2 and T_c/N^2 vs N for $\Delta O_c = 0.5$. As predicted by the theory and the numerics a crossover is found for groups for which opinions converge slower than the affinities: this is the signature of a distinctive difference in the behavior of small and large groups, numerically we found that this difference is effective for $N \sim 20$. Right inset: $< k >$ vs t/N^2 is plotted for two different values of N. As expected the two curves nicely collapse together.

5. Conclusion

In this paper we study a model of continuous opinions dynamics already proposed in[16] , which incorporates as main ingredient the affinity between agents both acting on the selection rule for the binary interactions as well entering the postulated mechanism for the update of the individual opinions.

Analyzing the model in the framework of adaptive networks we have been able to show that the sociological distinction between large and small groups can be seen as dynamical effect which spontaneously arises in our system. We have in fact proven that for a set of realistic parameters there exists a critical group size, which is surprisingly similar to the one reported in the psychological literature. Below this critical value agents first converge in mutual affinity and only subsequently achieved a final consensus on the debated issue. At variance, in large groups the opposite holds : The convergence in opinion is the driving force for the aggregation process, affinity converging on larger time scales.

Appendix A. Computation of γ_{eff}

The aim of this paragraph is to provide the necessary steps to decouple the opinion and affinity dynamics, by computing an effective value of the activating function $\Gamma_2(\Delta O_{ij}^t)$, hereby called γ_{eff}. This will be obtained by first averaging Γ_2 with respect to the opinions and then taking the time average of the resulting function:

$$\gamma_{eff} = \lim_{t \to \infty} \frac{1}{t} \int_0^t d\tau < \Gamma >_{op} (\tau), \qquad (A.1)$$

where the opinion–average is defined by:

$$< \Gamma >_{op} (t) = \int_0^1 dx \int_0^1 dy\, \Gamma_2(|x - y|) f(x) f(y),$$

being $f(\cdot)$ the opinions probability distribution function. To perform the computation we will assume that for each t, the opinions are uniformly distributed in the interval $[a(t), a(t) + L(t)]$ where $L(t) = 1 - t/T_c$ and T_c is the opinion convergence time, hence $f(\cdot) = 1/(NL(t))$. This assumption is motivated by the "triangle–like" convergence pattern as clearly dsplayed in the main panel of Fig. 1.

Assuming β_2 large enough, we can replace Γ_2 by a step function. Hence:

$$< \Gamma >_{op} (t) = \frac{1}{N^2 L^2(t)} \int_{a(t)}^{a(t)+L(t)} dx \int_{a(t)}^{a(t)+L(t)} dy\, \chi(x,y), \qquad (A.2)$$

where $\chi(x,y)$ is defined by (see also Fig. A1)

$$\chi(x,y) = \begin{cases} 1 & \text{if } |x-y| \ge \Delta O_c, \text{ i.e. in the triangles } T_1 \cup T_2 = Q \setminus D \\ -1 & \text{otherwise, i.e. in } D, \end{cases}$$
(A.3)

where Q is the square $[a, a+L] \times [a, a+L]$.

Let us observe that this applies only when $L(t) > \Delta O_c$ (see left panel of Fig. A1); while if $L(t) < \Delta O_c$ the whole integration domain, $[a, a+L] \times [a, a+L]$, is contained into the $|x-y| < \Delta O_c$ (see right panel of Fig. A1). In this latter case, the integration turns out to be simpler. In other words the

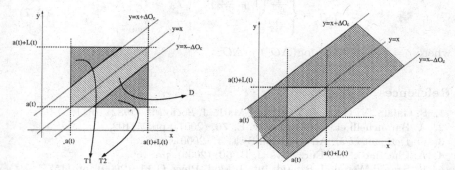

Fig. A1.   The geometry of the integration domains. On the left panel the case $L > \Delta O_c$, while on the right one the case $L < \Delta O_c$.

integral in Eq. (A.2) corresponds to measure the area, $|D|$, of the domains shown in Fig. A1 with a sign. Let us perform this computation according to the two cases: $L > \Delta O_c$ or $L < \Delta O_c$.

In the case $L > \Delta O_c$, the area of D is given by $|D| = |Q| - |T_1| - |T_2|$, hence

$$< \Gamma >_{op} (t) = -\frac{1}{N^2 L^2} (-|D| + |T_1| + |T_2|) = -\frac{1}{N^2 L^2} (-|Q| + 2|T_1| + 2|T_2|)$$

$$= -\frac{1}{N^2 L^2} \left(-L^2 + 4\frac{(L-\Delta O_c)^2}{2} \right) = \frac{1}{N^2} \left[1 - 2\left(1 - \frac{\Delta O_c}{L}\right)^2 \right] \quad \text{(if } L > \Delta O_c).$$
(A.4)

On the other hand if $L < \Delta O_c$, because the square Q is completely contained into the domain $|x-y| < \Delta O_c$ where χ is equal to -1, we easily get: $< \Gamma >_{op} (t) = -\frac{1}{N^2 L^2}(-L^2) = \frac{1}{N^2}$, if $L < \Delta O_c$. This last relation together with Eq. (A.4), can be casted in a single formula:

$$< \Gamma >_{op} (t) = \frac{1}{N^2} \left[1 - 2\left(1 - \frac{\Delta O_c}{L}\right)^2 \Theta\left(L - \Delta O_c\right) \right] ; \tag{A.5}$$

where Θ is the Heaviside function, $\Theta(x) = 1$ if $x > 0$ and zero otherwise. To conclude we need to compute the time average of $< \Gamma >_{op} (t)$. Using once again the "triangle–like"convergence assumption for the opinions, i.e. $L(t) = 1 - t/T_c$, where T_c is the opinion convergence time, we get:

$$\gamma_{eff} = \lim_{t \to \infty} \frac{1}{t} \int_0^t d\tau \frac{1}{N^2} \left[1 - 2 \left(1 - 2 \frac{\Delta O_c T_c}{T_c - \tau} + \left(\frac{\Delta O_c T_c}{T_c - \tau} \right)^2 \right) \Theta \left(\frac{T_c - \tau}{T_c} - \Delta O_c \right) \right],$$

(A.6)

This integral can be explicitly solved to give:

$$\gamma_{eff} = \begin{cases} \frac{1}{N^2} \left(1 + \frac{T_c \rho}{T_\alpha} \right) & \text{if } T_\alpha > T_c \\ \frac{\rho + 1}{N^2} & \text{if } T_\alpha < T_c, \end{cases}$$

(A.7)

where $\rho = -(1 + 2\Delta O_c \log(\Delta O_c) - \Delta O_c^2)$.

References

1. S. Galam, Y. Gefen & Y. Shapir, Math. J. Socio., **9**, (1982), pp.1.
2. A. Baronchelli et. al., Phys. Rev. E, **76**, (2007), pp.051102.
3. G. Deffuant et al. Adv. Compl. Syst. **3**, (2000), pp.87.
4. A. Pluchino et al., Eur. Phys. J. B, **50**, (2006), pp. 169.
5. K. Sznajd-Weron, J. Sznajd, Int. J. Mod. Phys. C **11**, (2000), pp.1157.
6. D. Stauffer & M. Sashimi, Physics A, **364**, (2006), pp.537.
7. W.R. Bion, Experiences in Groups, London: Tavistock (1961).
8. G.Le Bon, The Crowd: A Study of the Popular Mind, (1895); Repr. (2005).
9. W. McDougall, The Group Mind (1920).
10. R. Berk, A Collective Behavior, Dubuque, Iowa: Wm. C. Brown, (1974).
11. S.T. Fiske & S.L. Neuberg, in Advances in experimental social psychology Ed M. P. Zanna, Academic Press New York (1990).
12. S.L. Neuberg, J. of Personality and Social Psychology, **56** (1989), pp.374.
13. A. Bavelas, J. of Acoustical Sociology of America, **22**, (1950), pp.725.
14. A. Bavelas, Applied Anthropology, **7**, (1948), pp.16.
15. H.J. Leavitt, J. of Abnorm. and Soc. Psyc., **46**, (1951), pp.38.
16. F. Bagnoli et al., Phys. Rev. E, **76**, (2007), pp.066105.
17. T. Carletti et al., Euro. Phys. J. B, **64**, (2008), pp. 285.
18. T. Gross & B. Blasius, J. R. Soc. Interface, (2007),doi:10.1098/rsif.2007.1229.
19. M.G. Zimmermann, V.M. Eguíluz & M. San Miguel, Phys. Rev. E, **69**, (2004), pp.065102.
20. P. Holme & M.E.J. Newman, Phys. Rev. E, **74**, (2006), pp.056108.
21. B. Kozma & A. Barrat, Phys. Rev. E, **77**, (2008), pp.016102.
22. T. Carletti et al., preprint (2008).
23. T. Carletti et al., Europhys. Lett. **74**, (2), (2006), pp.222.

SHAPING OPINIONS IN A SOCIAL NETWORK

T. CARLETTI* and S. RIGHI

Département de Mathématique, Facultés Universitaires Notre-Dame de la Paix
Namur, Belgium, 5000
* E-mail: timoteo.carletti@fundp.ac.be
www.fundp.ac.be

We hereby propose a model of opinion dynamics where individuals update their beliefs because of interactions in acquaintances' group.

The model exhibit a non trivial behavior that we discuss as a function of the main involved parameters. Results are reported on the average number of opinion clusters and the time needed to form such clusters.

Keywords: sociophysics; opinion dynamics; agent based model; group interactions

1. Introduction

Complex Systems Science (CSS) studies the behavior of a wide range of phenomena, from physics to social sciences passing through biology just to mention few of them. The classical approach followed in the CSS consists first in a decomposition of the system into "elementary blocks" that will be successively individually analyzed in details, then the properties determined at micro–level are transported to the macro–level. This approach results very fruitful and shaped the CSS as an highly multidisciplinary field.

Recently models of opinion dynamics gathered a considerable amount of interest testified by the production of specialized reviews such as 1–3, reinforcing in this way the emergence of the *sociophysics*.[4] A basic distinction can be done in model of *continuous opinion with threshold*,[5,6] where opinions can assumed to be well described as continuous quantities; thus agents update their values because of binary interactions, if their opinions are close enough, i.e. below a given threshold. The second class consists of models where opinions can be described by *discrete* variable, yes/no for instance, and they are updated according to *local rules, i.e. small group interactions* as for instance: majority rule,[7] majority and inflexible rules,[8] majority and contrarian rules.[9]

In this paper we introduce a new model of opinion dynamics which naturally sets at the intersection of the former scheme; in fact individuals have continuous opinions that are updated if they are below some given threshold, once agents belong to a group, whose size evolves dynamically together with the opinion. Moreover each agent possesses an *affinity* with respect to any other agent, the higher is the affinity score the more trustable is the relationship. Such affinity evolves in time because of the past interactions, hence the acquaintances' group is determined by the underlying evolving social network.

We hereby provide an application of the model to the study of the consensus–polarization transition that can occur in real population when people do agree on the same idea – consensus state – or they divide into several opinion groups – polarization state.

The paper is organized as follows. In the next section we introduce the model and the basic involved parameters, then we present the results concerning the consensus–polarization issue and the time needed to reach such asymptotic state. We will end with some conclusions and perspectives.

2. The model

The model hereby studied is a generalization of the one proposed in[6,10] because now interactions occur in many–agents groups, whose size is not fixed a priori but evolves in time.

We are thus considering a fixed population made of N agents, i.e. closed group setting, where each agent is characterized by its opinion on a given subject, here represented by a real number $O_i^t \in [0,1]$, and moreover each agent possesses an affinity with respect to any other, $\alpha_{ij}^t \in [0,1]$: the higher is α_{ij}^t the more affine, say trustable, the relationships are and consequently agents behave.

At each time step a first agent, say i, is randomly drawn with an uniform probability, from the population; then, in order to determine its acquaintances' group, it computes its *social distance* with respect to the whole population:

$$d_{ij}^t = |O_i^t - O_j^t| \left(1 - \alpha_{ij}^t\right), \forall j \in \{1, \ldots, N\} \setminus \{i\}. \tag{1}$$

The agents j whose distance from i is lesser than a given threshold, Δg_c, will determine the acquaintances' group of i at time t in formula:

$$\mathcal{F}_i^t = \left\{j : d_{ij}^t \leq \Delta g_c\right\}. \tag{2}$$

Let us observe that the group changes in time, in size and in composition, because the opinions and/or affinities also evolve. The rationale in the use of the affinities in the definition of the social metric is to interpret[10,11] the affinity as the adjacency matrix of the (weighted) social network underlying the population. We hereby assume a constant threshold Δg_c for the whole population, but of course one could consider non–homogeneous cases as well.

Once the agent i has been selected and the group $\mathcal{F}_i^t$ has been formed, the involved individuals do interact by possibly updating their opinions and/or affinities. Once in the group, all agents are supposed to listen to and speak to all other agents, therefore every one can perceive a personal *averaged* – by the mutual affinity – *group opinion*, $< O_l^t >$, because each agent weights differently opinions of trustable agents from the others. In formula:

$$< O_l^t >= \frac{\sum_{j=1}^{m_i} \alpha_{lj}^t O_j^t}{\sum_{j=1}^{m_i} \alpha_{lj}^t} \quad \forall l \in \mathcal{F}_i^t, \tag{3}$$

where we denoted by m_i the size of the subgroup $\mathcal{F}_i^t$. The vector $(< O_1^t >$ $, \ldots, < O_{m_i}^t >)$, will hereby named *apopsicentre*, i.e. the barycentre of the opinions ($\acute{\alpha}\pi o\psi\eta$ = opinion).

Because the affinity is is general not symmetric, some agents could have been included in the group determined by i, "against" their advise, hence a second relevant variable is the *averaged affinity* that each agent perceives of the group itself [a] :

$$< \alpha_l^t >= \frac{1}{m_i} \sum_{j=1}^{m_i} \alpha_{lj}^t \quad \forall l \in \mathcal{F}_i^t. \tag{4}$$

Once the former two quantities have been computed by each agent, we propose the following update scheme: to belong in the largest size group, each agent, would like to come closer to its perceived apopsicentre if it feels himself affine enough to the group:

$$O_l^{t+1} = O_l^t + \frac{1}{2} \left(< O_l^t > - O_l^t \right) \Gamma_1(< \alpha_l^t >) \quad \forall l \in \mathcal{F}_i^t, \tag{5}$$

where $\Gamma_1(x) = \frac{1}{2} \left[\tanh(\beta_1(x - \alpha_c)) + 1 \right]$ is an activating function, defining the region of trust for effective social interactions, e.g. $< \alpha_l^t >$ larger than α_c.

---

[a] Each agent in the group can determine the apopsicentre, but if it is in an hostile group, it will not move toward this value.

Moreover sharing a close opinion, reinforce the mutual affinity, while too far opinions make the relationship to weak, hence each agent becomes more affine with all the agents in the subgroup that share opinions close enough to its perceived apopsicentre, otherwise their affinities will decrease:

$$\alpha_{jk}^{t+1} = \alpha_{jk}^t + \alpha_{jk}^t \left(1 - \alpha_{jk}^t\right) \Gamma_2(\Delta O_{jk}) \quad \forall j, k \in \mathcal{F}_i^t, \tag{6}$$

where $\Delta O_{jk} \doteq <O_j^t> - O_k^t$, and

$$\Gamma_2(x) = \tanh\left[\beta_2\left(\Delta O_c - |x|\right)\right], \tag{7}$$

that can be considered again as an activating function for the affinity evolution. In the previous relations for $\Gamma_{1,2}$, we set the parameters β_1 and β_2, large enough to practically replace the hyperbolic tangent with a simpler step function. Under these assumptions Γ_1 takes values either 0 or 1, while the value of Γ_2 are either -1 or 1. The interaction mechanism is schematically represented in Fig. 1.

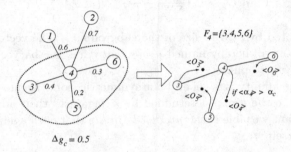

Fig. 1. Cartoon to represent the group formation and the interaction mechanism. On the left panel, the composition of the local acquaintance group. On the right panel, dynamics in the opinion space : each agent tends to move following the represented arrows.

3. Results

A typical run of this model is presented in Fig. 2. First of we can observe that the dynamics is faster than in the similar model presented in 6,10, this is because binary interactions are replaced by multi–agents interactions that improve the information spread. Moreover there exists a transient interval of time, where nobody modifies its opinion, but only the mutual affinities (see insets of Fig. 2 and the relative caption). Only when the relationships become highly trustable, agents do modify also their opinions.

This behavior is explained by the different time scales of the main processes: evolution of opinions and evolution of affinity, as it clearly emerges again from the insets of Fig. 2, where we show three time-snapshots of the affinity, i.e. the social network, once agents start to modify their opinions. Transferring this behavior to similar real social experiment, we could expect that, initially unknown people first change (in fact construct) their affinities relationships (increasing or decreasing mutual affinities) and only after that, they will eventually modify their opinions. Namely initially people "sample" the group and only after they modify their beliefs.

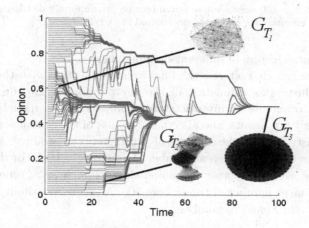

Fig. 2. Time evolution of the opinion (main panel) and the social network of affinity time-snapshots (small insets). Parameters are: $\Delta g_c = 0.1$, $\alpha_c = 0.5$, $\Delta O_c = 0.5$, $N = 100$ agents whose opinion are initially uniformly distributed in $[0, 1]$, whereas initial affinities are uniformly distributed in $[0, 1/2]$. Three time-snapshot of social networks are reported for increasing times, $T_3 > T_2 > T_1$. Dots represent agents that are possibly linked if they are affine enough. Each network has been characterized by some standard topological indicator; the diameter and the averaged shortest path (that are infinite for G_{T_1} and respectively have values 4 and 1.81 for G_{T_2} and 2 and 1.00 for G_{T_3}), the averaged degree (that takes values 0.07, 0.39 and 0.9 respectively for G_{T_1}, G_{T_2} and G_{T_3}) and the averaged network clustering (that assumes the values 0.10, 0.72 and 0.99 respectively for G_{T_1}, G_{T_2} and G_{T_3}).

In Fig. 3 we report two different outcomes of numerical simulations of the model, for two sets of parameters, in the left panel we found once again a consensus status, where all the population share the same opinion, as reported by the histogram. While in the right panel, the population

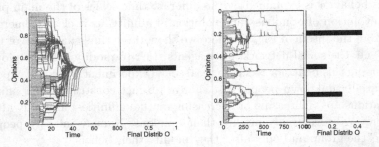

Fig. 3. Time evolution of the opinions and asymptotic opinion distribution. Parameters are: $\Delta g_c = 0.1$, $\alpha_c = 0.5$ (left panel), and $\Delta g_c = 0.02$, $\alpha_c = 0.5$ (right panel) both with $\Delta O_c = 0.5$ and $N = 100$ agents whose initial opinion are uniformly distributed in $[0, 1]$, whereas initial affinities are uniformly distributed in $[0, 1/2]$.

polarizes [b] into clusters of different opinions, here 4.

Hence one can characterize the final asymptotic state with the number of opinion clusters as a function of the key parameters Δg_c and α_c. We observe that often the asymptotic state exhibits *outliers*, namely clusters formed by very few agents, and also that, because of the random encounters and initial distributions, the same set of parameters can produce asymptotic state that can exhibit a different number of opinion clusters. For this reason we define the *average number* of opinion clusters, $< N_{clu} >$, repeating the simulation a large number of times, here 500. A second possibility is to use the *Deridda and Flyvbjerg* number:[12]

$$Y = \sum_{i=1}^{M} \frac{S_i^2}{N^2}, \qquad (8)$$

where M is the total number of clusters obtained in the asymptotic state and S_i is the number of agents in the i-th cluster. The quadratic dependence on S_i/N ensures that less weight has been given to small clusters with respect to larger ones.

---

[b]Let us observe that polarized case might be metastable; in fact if the mean separation between the adjacent opinion peaks is smaller than the opinion interaction threshold, ΔO_c, and Δg_c is not too small, there always exists a finite, though small, probability of selecting in the same acquaintance group individuals belonging to different opinion clusters, hence producing a gradual increase in the mutual affinities, which eventually lead to a merging of the, previously, separated clusters. This final state will be achieved on extremely long time scales, diverging with the group size: socially relevant dynamics are hence likely to correspond to the metastable regimes. A similar phenomenon has been observed in 6,10.

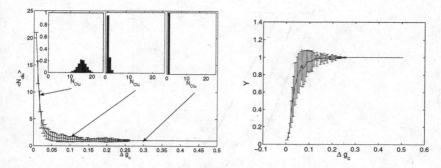

Fig. 4. Number of clusters (left panel) and Derrida and Flyvbjerg (right panel) as function of Δg_c when $\alpha_c = 0.5$. Average and Standard Deviation on 200 simulations. The distributions of the average number of cluster are presented for $\Delta g_c = 0.3$ (right inset), $\Delta g_c = 0.1$ (central inset) and $\Delta g_c = 0.01$ (left inset).

In Fig. 4 we report the results of the analysis of $< N_{clu} >$ and Y as a function of Δg_c, for a fixed value of α_c.

A phase transition from a mono–cluster state, i.e. *consensus*, to polarization of the opinions in a population of 100 agents, emerges close to $\Delta g_c = 0.25$, for smaller values of Δg_c the distribution of the number of cluster can be well described by a normal distribution (see left inset Fig. 4), for larger value of Δg_c, only one cluster is present (see right inset Fig. 4), while for Δg_c varying around 0.25 an exponential distribution can be found (see middle inset Fig. 4), reinforcing thus the statement of the existence of a phase transition.

Data from Fig. 4 suggest an exponential growth of $< N_{clu} >$ as a function of Δg_c below the phase transition value, we thus compute a linear fit on log–log scale (see Fig. 5) in the region involving small values of Δg_c, obtaining:

$$\log < N_{clu} >= -1.495 \log(\Delta g_c) - 4.107 \,,$$

when $\alpha_c = 0.5$. A similar power law behavior is still valid also for the Derrida and Flyvbjerg number. The existence of a power low seems robust with respect to variations of the parameter α_c (see Fig. 5). The results presented in Fig. 5 allow us to extract also the behavior of the average number of clusters as a function of the second parameter α_c for a fixed Δg_c. In fact moving upward on vertical lines, i.e. decreasing α_c, the $< N_{clu} >$ increases if Δg_c is below the critical threshold, while above this value the number of clusters is always equal to one. Moreover from these data we could conclude that the phase transition point seems to be independent from the value of α_c.

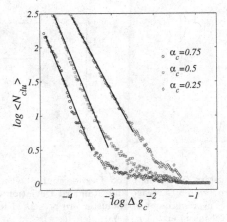

Fig. 5. Average number of clusters as function of Δg_c (log–log scale). Best linear fits for different values of α_c. Each simulation has been repeated 200 times.

Another relevant quantity of interest, is the time needed to form an opinion cluster, the *opinion convergence time*, T_c, and its dependence of the size of the cluster. Numerical simulations not reported here, emphasize that in the polarization case T_c depends in a highly non–trivial way on the total number of clusters and on their sizes, roughly speaking if in consensus state the time needed to form a cluster of say N_1 individual is some value T_1, then the time needed to form a cluster of the same size in a polarization case, has nothing to do with T_1 and it depends on all the formed clusters.

Because this paper offers a preliminary analysis, we decided to consider only the consensus case, hence choosing parameters ensuring the existence of only one cluster and we define the *convergence time* T_c to be:

$$T_c = \min \left\{ t \geq 0 : \max_i(O_i(t)) - \min_i(O_i(t)) \leq \frac{a}{N} \right\}, \tag{9}$$

where a is a small parameter (hereby $a = 0.1$).

We thus performed dedicated simulations with N ranging from few unities to thousand unities. The results reported in Fig. 6 suggest a non–linear dependence of T_c on N, well approximable by $T_c \sim N^b$. Using a regression analysis on the data, we can estimate the exponent which results, $b = 0.091$ for $\Delta g_c = 0.6$ and $b = 0.087$ for $\Delta g_c = 0.5$. Let us observe that as Δg_c approaches the phase transition value, T_c increases and the curve becomes more noisy (see for instance the bottom panel of Fig. 6 corresponding to $\Delta g_c = 0.4$), that is because the occurrence of the consensus case becomes lesser and lesser probable.

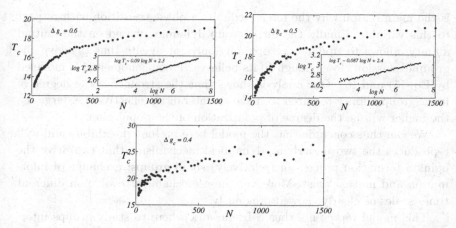

Fig. 6. Time of convergence of the opinions as function of number of agents. Panels correspond to parameters $\Delta g_c = 0.6$ (top-left), $\Delta g_c = 0.5$ (top-right) and $\Delta g_c = 0.4$ (bottom). Insets log–log plots of T_c as a function of N.

4. Conclusions

In this paper we introduced a new model of opinion dynamics where agents meet in social groups, affinity driven, and possibly update their beliefs as a consequence of these local, i.e. group level, interactions. The model exhibits a rich phenomenology determined by the interlay between the dynamics of the opinions and the mutual affinities. We emphasized the role of two parameters, Δg_c and α_c, which can be qualitatively interpreted respectively as the openness of mind in the formation of the group and as the openness of mind in the intra–group dynamics. We thus studied the behavior of the model as a function of these two parameters.

The formulation of our model has been inspired by the observation of the way in which the formation mechanisms for social interactions do occur in the real world: a large majority of the processes of formation and evolution of the opinions are driven by group based discussions, such groups are determined by the mutual affinity and/or the shared opinion. The processes of group formation which tends to form clusters of acquaintances (or collaborators) are introduced in our model via the selection mechanism based on the mutual trust, i.e. Eq. (1) and (2).

The numerical analysis we performed, shows a dependence of the consensus/polarization state on the degree of mind openness in the creation of the groups of acquaintances, i.e. the parameter Δg_c, large values corresponding to consensus states while small ones to fragmented groups. That

234

is the main reason why the model exhibits a phase transition with respect to this variable. Finally the intra–group dynamics, based on the mutual affinity, allows to update the opinions only for agents that perceive the group discussion close enough to its believes. This phenomenon is modeled by Eq. (5) and (6). Our analysis shows that the stronger is the degree of intra–group affinity required to make an interaction effective, i.e. large α_c, the higher will be the degree of polarization of the population.

We can thus conclude that the model here presented, exhibits and well reproduces the two underlying dynamical mechanisms that can drive the opinion formation process in (relatively) small groups: exchange of information and mutual trust. Moreover these mechanisms evolve on different times scales as clearly showed previously.

This model represents thus a framework where to study groups interactions with applications to real social systems. It would be interesting to improve the model by introducing, at least, two factors: vectorial opinions[3,13,14] i.e. agents discuss and exchange information about more than one subject. Second, introduce a limitation in the number of agents with which anyone can be affine with, as usually is the case in the real social networks.

References

1. S. Galam, arXiv:physics.sco-ph/0803.1800v1, (2008).
2. R. Hegselmann and U. Krause, JASSS, **5**, no.3, (2002).
3. C. Castellano et al., preprint arXiv: 0710.3256, submitted Review of Modern Physics, (2008).
4. S. Galam and Y. Shapir, Math. J. Socio., **9**, (1982), pp. 1.
5. G. Weisbuch et al, Complexity, **7**, No. 3, 2002.
6. F. Bagnoli et al., Phys. Rev. E, **76**, (2007), pp.066105.
7. S. Galam, Europhys. Lett., **70**, (6), (2005), pp. 705.
8. S. Galam and F. Jacobs, Physica A, **381**, (2007), pp. 366.
9. S. Galam, Physica A, **333**, (2004), pp. 453.
10. T. Carletti et al., accepted proceedings WIVACE2008, (2008).
11. T. Carletti et al., Euro. Phys. J. B, in press, (2008).
12. B. Derrida and H. Flyvbjerg, J. Phys. A, **19**, (1986), pp. 1003.
13. R. Axelrod, J. Conflict Resolution, 41, (1997), pp. 203.
14. S. Fortunato et Al., Int. Jour. Mod. Phys. C, 16, (2005),pp. 1535.

A CELLULAR AUTOMATA MODEL FOR HIGHWAY TRAFFIC WITH PRELIMINARY RESULTS

S. DI GREGORIO and R. UMETON*

Department of Mathematics, University of Calabria
Arcavacata, Rende, CS, 87036, Italy.
www.unical.it

A. BICOCCHI and A. EVANGELISTI

Abstraqt srl, via Puccini 311
Lucca, LU, 55100, Italy.

Cellular Automata are an established formal support for modelling traffic. STRATUNA is a Cellular Automata model for simulating two/three lanes highway traffic. It is based on an extensive specification of the driver response to the surrounding conditions. The model is deterministic with regard to driver behaviour, even if values of parameters ruling the reactivity level of the drivers are assigned stochastically. Probability distribution functions were deduced by field data and applied to vehicular flow generation (vehicle types, driver desired speed, entrance-exit gates). A partial implementation of STRATUNA was performed and applied to Italian highway A4 from Venice to Trieste. Simulations were compared with available field data with results that may be certainly considered encouraging in this initial implementation.

Keywords: cellular automata; modelling and simulation; highway traffic.

1. Introduction

Cellular Automata (CA) are a computational paradigm for modelling high complexity systems[1] which evolve mostly according to the local interactions of their constituent parts (acentrism property). Intuitively a CA can be seen as a d-dimensional space, partitioned into cells of uniform size, each embedding a computational device, the elementary automaton (EA), whose output corresponds to its state. Input for each EA is given by states of EA in neighbouring cells, where neighbouring conditions are determined by a

---
*Corresponding author. Tel.: +39 0984 496467; fax: +39 0984 496410.
Email addresses: dig@unical.it (S. Di Gregorio), umeton@mat.unical.it (R. Umeton).

pattern invariant in the time and equal for each cell. EA are in an arbitrary state at first (initial conditions), subsequently CA evolves by changing simultaneously states to all the EA at equal discrete time steps, according to EA transition function (parallelism property).

CA were used for modelling highway traffic[2] because of acentric and parallel characteristics of such a phenomenon. As a matter of fact, when highway structural features are fixed and there are no external interferences out of the vehicular interactions (normal conditions), the traffic evolution emerges by the mutual influences among vehicles in driver sight range.

The main CA models of highway traffic[3-6] (to our knowledge) may be considered "simple" in terms of external stimuli to the driver and corresponding reactions, but they are able to reproduce the basic three different phases of traffic flow (i.e. free flow, wide moving jams and synchronized flow) by simulations to be compared with data (usually collected automatically by stationary inductive loops on highways).

We developed STRATUNA (Simulation of highway TRAffic TUNed-up by cellular Automata), a new CA model for highway traffic with the aim to describe more accurately driver surrounding conditions and responses. We referred to a previous CA model[7,8] , that was enough satisfying in the past, but now it is dated for the different technological situations (e.g., the classification of vehicles on the base of pure acceleration, deceleration features is no more realistic). Reference data for deducing STRATUNA parameters and for real-simulated event comparison are the timed highway entrance-exit data, that are comprehensive of the vehicle type.

The paper's next section outlines the STRATUNA model, while the transition function is described in the third section. Partial implementation of the model is discussed together with simulation results and comparison with real event in the fourth section. Conclusions are reported at the end.

2. The STRATUNA general model

STRATUNA is based on a "macroscopic" extension of CA definition[1] , involving "substates" and "external influences". The set of "state values" is specified by the Cartesian product of sets of "substate values". Each substate represents a cell feature and, in turn, a substate could be specified by sub-substates and so on. Vehicular flows at tollgates and weather conditions are external influences, generated by dataset or probabilistic functions according to field data and are applied before the CA transition function.

Only one-way highway traffic is modelled by STRATUNA (complete highway is obtained by a trivial duplication). One-dimension is sufficient, because a cell is a highway segment, 5m long, whose specifications (sub-states) enclose width, slope and curvature in addition to features of possible pairs vehicle-driver. The STRATUNA time step, the driver minimum reaction time, may range from 0.5s to 1s (CA clock).

An 8-tuple defines $STRATUNA = \langle R, E, X, P, S, \mu, \gamma, \tau \rangle$, where:

- $R = \{x | x \in \mathbb{N}, 1 \leq x \leq n\}$ is the set of n cells, forming the highway.
- $E \subseteq R$ is the set of entrance-exit cells in R, where vehicles are generated and annihilated.
- $X = \langle -b, -b+1, ..., 0, 1, ..., f \rangle$ defines the EA neighbouring, i.e the forward (f) cells and backward (b) cells in the driver sight, when visibility is maximum (no cloud, sunlight etc.).
- $P = \{length, width, clock, lanes\}$ is the set of global parameters, where $length$ is the cell length, $width$ is the cell width, $clock$ is the CA clock, $lanes$ is the number of highway lanes (1, 2 .. from right to left), that includes an additional lane 0, representing from time to time the entrance, exit, emergency lane.
- $S = Static \times Dynamic \times (Vehicle \times Driver)^{lanes}$ specifies the high level EA substates, that are clustered in typologies, i.e statical and dynamical features of highway segment corresponding to the cell, vehicle and driver features (there are at most as many pairs vehicle-driver as lanes). Such substates are detailed in the Table 1.
- $\mu : \mathbb{N} \times R \rightarrow Dynamic$ is the "weather evolution" function, that determines $Dynamic$ values for each step $s \in \mathbb{N}$ and each cell $c \in R$.
- $\gamma : \mathbb{N} \times E \rightarrow Vehicle \times Driver$ is the vehicle-driver pair $normal$ generation function for each step $s \in \mathbb{N}$ and each cell $c \in E$.
- $\tau : S^{b+1+f} \rightarrow S$ is the EA transition function. The visibility reduction to b' backward cells and to f' forward cells involves that cells out of range will be considered without information.

Table 1. Substates and related sub-substates.

Substate name	Sub-substates self-explanatory names
Static	*CellNO, Slope, CurvatureRadius, SurfaceType, SpeedLimit, LaneISpeedLimit*
Dynamic	*BackwardVisibility, ForwardVisibility, Temperature, SurfaceWetness, WindDirection, WindSpeed*
Vehicle	*Type, Length, MaxSpeed, MaxAcceleration, MaxDeceleration; CurrentSpeed, CurrentAcceleration, Xposition, Yposition, Indicator, StopLights, WarningSignal*
Driver	*Origin, Destination, DesiredSpeed, PerceptionLevel, Reactivity, Aggressiveness*

3. The STRATUNA transition function

An overview of the transition function will be here given with the aim of exposing the leading ideas and the adopted choices concerning STRATUNA, together with a better specification of the mentioned substates and sub-substates.

A vehicle is specified by constant and variable (during all the simulation) values of sub-substates. Constant properties are *Type* (motorcycle, car, bus/lorries/vans, semitrailers/articulated), *Length*, *MaxSpeed*, *MaxAcceleration*, *MaxDeceleration*. The main mechanism of the traffic evolution is related to the determination of the new values of the variable sub-substates of *Vehicle*, i. e., *Xposition* and *Yposition* (they individuate the cell co-ordinates x, y of the middle point in the vehicle front) *CurrentSpeed*, *CurrentAcceleration*, *Indicator* (with values: null, left, right, hazard lights) *StopLights* (on, off), *WarningSignal* (on, off).

Note that the vehicle space location is not identified by a sequence of full cells as in other CA models[2] , but it is more accurate because portions of cell and positions between two lanes can be considered occupied. *Indicator* and *WarningSignal* sub-substates in the simulation hold a larger role than indicator and a generic warning signal in the real events. When a real driver wants to change lane, not always he uses the indicator, but drivers around detect such a manoeuvre from his behaviour (e.g. a short beginning moving toward the new lane before to decide overtaking). Of course simulation doesn't account for these particular situations, but this problem doesn't exist, a driver in the simulation communicates his intention to change lane always by the indicator. Sub-substate *WarningSignal* is activated when driver wants to signal that he needs the lane immediately ahead of his vehicle to be free. This situation corresponds in the real word to different actions or their combination, e.g. sounding the horn, blinking high-beam lights, reducing "roughly" the distance with vehicle ahead and so on. Through such sub-substates, *Indicator*, *StopLights*, *WarningSignal* a communication protocol could be started between vehicles.

The single vehicle V moving involves two computations, i.e. the objective determination of the future positions of vehicles "around V" and the subjective V driver reaction. The former one is related to the objective situation and forecasts all the spectrum of possible motions of all the vehicles, that can potentially interact with V, i.e. the vehicles in the same cells, where V extends more the next vehicles ahead and behind such cells for each lane in the range of the neighbourhood.

In first instance, some *Static* and *Dynamic* sub-substates determine highway conditions (e.g. highway *surface_slipperiness* is computed by *SurfaceType*, *SurfaceWetness* and *Temperature*); subsequently, they are related to the *Vehicle* sub-substates in order to determine the temporary variable *max_speed* that guarantees security with reference only to the conditions of highway segment represented by cell. It accounts for the vehicle stability, speed reduction by limited visibility and speed limits in the lane, occupied by the vehicle. If *max_speed* is smaller than *DesiredSpeed*, *desired_speed* = *max_speed* otherwise *desired_speed* = *DesiredSpeed*. *Slope* and *surface_slipperiness* determine the temporary variables *max_acceleration* and *max_deceleration*, correction to sub-substates *MaxAcceleration*, *MaxDeceleration*.

The next computation step determines "objectively" the "free zones" for V, i.e. all the zones in the different lanes, that cannot be occupied by the vehicles around V, considering the range of the speed potential variations and the lane change possibility, that is always signalled by *Indicator*. Note that the possible deceleration is computed on the value of *max_deceleration* in the case of active *StopLights*, otherwise a smaller value is considered, because deceleration could be only obtained by shift into a lower gear or by relaxing the accelerator.

The last computation step involves the driver subjectivity. First of all, the cell number corresponding to vehicle position *CellNO* is compared with the cell number of *Destination* in order to evaluate if the exit is so close to force approaching lane 1 (if in other lanes) or continuing in lane 1 slowing down opportunely to the ramp speed limit.

The driver aims in the other cases to reach/maintain the *desired_speed*; different options are perceived available, each one is constituted by actions (Fig. 1) involving costs (e.g. the cost of the gap between the new value of *CurrentSpeed* and the *desired_speed*). The driver chooses the option, among all the possible ones, with minimal sum of the costs.

All is based on a driver subjective perception and evaluation of an objective situation by sub-substates *PerceptionLevel*, *Reactivity*, *Aggressiveness*. *PerceptionLevel* concerns the perception of the free zones; their widths are reduced or (a little bit) increased by a percentage before to compute on their new values the various possibilities to reach free zones in security conditions, considering the variable values of *Vehicle* sub-substates more *max_speed*, *max_acceleration* and *max_deceleration*.

Reactivity is a collection of constants for determining costs by means of function of the same type expressed in Fig. 1. Examples are "remaining in a takeover lane", "staying far from *desired_speed*", "breaking significantly", "starting a takeover protocol".

Aggressiveness forces the deadlocks, that could be generated by a cautious *PerceptionLevel*, e.g. when the entrance manoeuvre is prohibited in a busy highway, because free zones are very much reduced in the perception phase. The stop condition increases at each step the *Aggressiveness* value, it implies a proportional increase of the percentage value of *PerceptionLevel* from negative values to positive ones until the free zone in a lane remains shorter than the distance between two consecutive vehicles, where the entrance could be performed. *Aggressiveness* value comes back to zero when stop condition ends.

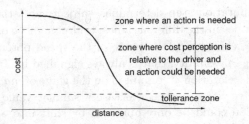

Fig. 1.   The function that connects the distance from front vehicle with a cost.

4. STRATUNA first partial implementation

At present, STRATUNA has been partially implemented in a simplified form in order to perform a preliminary validation. The implemented model is the $\beta4$ version: $STRATUNA_{\beta4} = \langle R, E, X', P, S', \gamma_{\beta4}, \tau_{\beta4} \rangle$.

The function μ disappeared, because no weather evolution is considered, but only constant average conditions. Therefore $X' = \langle -r, -r + 1, ..., 0, 1, ..., r \rangle$ substitutes X where r is a radius, accounting for the average visibility of an average driver and *Dynamic* substate is no more considered. *Indicator* lacks of hazard lights value, *PerceptionLevel* value is always 1, behaviour involving *Aggressiveness* was not implemented and *Reactivity* is considered only for "staying far from *desired_speed*".

The generation function $\gamma_{\beta4}$, was tailored for the traffic of Italian highway A4 Venezia-est/Lisert, characterized by two lanes and twelve entrances/exits. Data are composed by around 1 milion of tolltickets, they are related to 5 non-contiguous weeks and grouped in five categories, depending on vehicle number of axles (it is reducible to our vehicle classification). Due to problems of time synchronization among tollgates, these datasets have to be considered partial and incomplete. For these reasons, a data cleaning step was mandatory for the following infrequent situations: (i) missed tickets: transits without entrance or starting time; (ii) transits across two or more days; (iii) transits that end before beginning; (iv) vehicles too fast to be true: exceeding 200 km/h as average speed. Afterwards, the average speed was related to the total flow for each of the 34 days.

The result of this quantitative study is summarized in the following chart: each day is rappresented as a dot; a shift over x-axis and y-axis is a variation respectively of "total flow" and "average speed" from their averaged values over all the days (Fig. 2a). *DesiredSpeed* distribution (Fig. 2b) according to the vehicle *Type* are easily deduced by highway data in free flow conditions for vehicles covering short distance in highway. The probability to park in the rest and services areas is minimal in short distance cases. Parking in the rest and services areas cannot be detected by data and causes errors; they justify the slightly higher values of average speed obtained in the simulated cases, in comparison to the same values of corresponding real events. Finally a statistical sampling treatment was performed to select meaningful subsets. After scaling flow values and vehicle generation rate, some validation sets were designed. Each set provides a number of vehicles (each one specified by the couple $\langle Origin, Destination \rangle$) and the average real speed ($\overline{rS}$) over all its vehicles and over all the event. Being 95% of real traffic, generated vehicles are all cars. Validation sets concern conditions from freeflow to congestion situation. In order to give a recapitulation of salient characteristics of implemented transition function, a pseudo-code block is proposed in appendix.

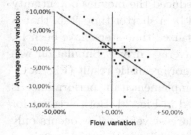

Vehicle type	Desired speed	Flow share
I	122.80 km/h	93.4%
II	112.77 km/h	4.6%
III	113.29 km/h	0.5%
IV	102.61 km/h	0.1%
V	93.90 km/h	1.4%

(a) Daily flow and speed fluctuation from the average

(b) Share and desired speed for each type of vehicle in selected case of freeflow

Fig. 2.   Daily and selected data.

242

4.1. *Results of simulations with* $STRATUNA_{\beta 4}$

We report five significant simulations for typical highway conditions: freeflow (Fig. 3a), moderated-flow next to congestion (Fig. 3b and Fig. 3c) and locally congested situations (Fig. 3d). In addition to $\overline{rS}$ (represented in figures as a line) we consider step-by-step average simulated speed ($\overline{sS}$, represented in figures as fluctuating curves) and average simulated desired speed ($\overline{sDS}$, represented in figures as an invariant notch, Cfr. Fig. 2b). Simulation conditions contemplate, at the beginning, for all the cases, an empty highway, fed at each entrance with vehicles according to appropriate generation rate. Initially, average speed is low, because generated vehicles start from null speed. After this very first phase, $\overline{sS}$ increases since vehicles can tend to their *DesiredSpeed* value, until the small number of vehicles in the highway permits free flow conditions (i.e. when simulation time < 500s). To provide a goodness measurement, simulations reported are accompanied with two error quantification: e_1 and e_2. The first one measures the average relative error (over all AC steps) between $\overline{sS}$ and $\overline{rS}$; the second one is the same as first but calculated after 500 seconds of simulated time in order to skip the initial phases of model evolution.

In the freeflow case, $\overline{sS}$ matches $\overline{rS}$ during the whole simulation, remaining slightly higher than field data, with very short oscillations (Fig. 3a). In the moderated flow case (Fig. 3b), after the same initial phase, $\overline{sS}$ became definitely lower than $\overline{rS}$ with moderate oscillations. Such a behaviour is not correct, also if its error rate is low: the cars in the simulation must be faster than corresponding real cars, because they don't waste time to park in the rest and services areas. This problem depends clearly on the driver subjective evaluation, that came out too much cautiously because the partial implementation of transition function reduced the moving potentiality (reaction rigidity). A possible solution could be a shorter time step, that is equivalent to a more rapid reactivity. The utilized time steps have been 1s, the standard average reaction time of the average driver. Simulation was repeated with time step 0.75s, obtaining a more realistic result (Fig. 3c).

After this, two simulations, where the implementation performance is lower than previous simulations, are reported in Fig. 3d. Both account for the same particular real situation, when a largest vehicle flow occurs only from one entrance; both run on the same model specifications and feed function. $\overline{sS}$ became quickly significantly higher than $\overline{rS}$. This means that the reaction rigidity of the driver was rewarded by a higher speed in this particular case because the entrance filtering creates synchronization.

Classical patterns of highway traffic (moving jams and synchronized

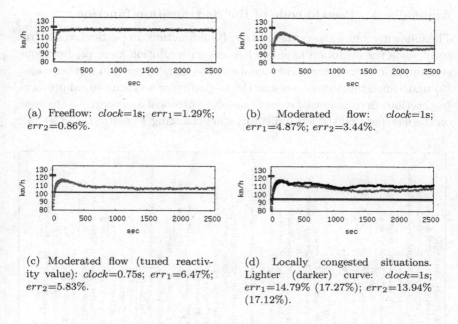

(a) Freeflow: $clock$=1s; err_1=1.29%; err_2=0.86%.

(b) Moderated flow: $clock$=1s; err_1=4.87%; err_2=3.44%.

(c) Moderated flow (tuned reactivity value): $clock$=0.75s; err_1=6.47%; err_2=5.83%.

(d) Locally congested situations. Lighter (darker) curve: $clock$=1s; err_1=14.79% (17.27%); err_2=13.94% (17.12%).

Fig. 3.   Average speed fluctuation in selected case study.

flow) were observed in the simulations of congested traffic, but the lack of data collected automatically by stationary inductive loops (single vehicle data[2]) doesn't permit a serious comparison.

5. Conclusions

Preliminary results of the reduced version $\beta 4$ of the STRATUNA model are very promising, considering that discrepancies between statistics deduced by real data and simulations are in part justified by unavoidable inaccuracies in the available real data and by imprecision introduced by parking in the rest and services areas. This is an interesting starting point in order to implement the full model. An important problem will be to tune some values of variables concerning the driver subjective behaviour to solve problems of congested situations. Such values cannot be deduced directly by data of real events as the *DesiredSpeed*, therefore a calibration action must be performed. We intend to use the powerful tool of Genetic Algorithms in order to solve this crucial problem. Accessing to other types of data concerning highway traffic would be important for the approach completeness.

Appendix A. Pseudo-code of the $\beta 4$ transition function

The following figure shows the $\beta 4$ transition function. In particular, (1) "return" ends the evolution of single EA at each evolution step; (2) functions starting in lowercase are actions enqueued to be performed in further steps; (3) underlined functions represent the beginning of a synchronized protocol (e.g. actions in consecutive steps of takeover-protocol are: control a freezone on the left, light on the left indicator, start changing $Y position$, and so on).

```
BEGIN: TransitionFunction()
FindNeighbours(); ComputeSpeedLimits();
ComputeTargetSpeed(); DefineFreeZones();
AssignTheCost_PM_WhereAFreeZoneIsReduced();
if(ManouvreInProgress())
  continueTheManouvre(); return;
if(myLane==0) //I'm on a ramp
  if(IWantToGetIn())
    if(ICanEnter())
      enter(); return;
    else
      if(IHaveSpaceProblemsForward())
        slowDown(); return;
      else followTheQueue(); return;
  else //the ramp is not ended yet
    if(IHaveSpaceProblemsForward())
      followTheQueue(); return;
    else keepConstantSpeed(); return;
  else // I want to get out
    if(TheRampEnded()) deleteVehicle(); return;
    else
      if(IHaveSpaceProblemsForward())
        followTheQueue(); return;
      else keepConstantSpeed(); return;
//end lane==0
else if(myLane==1)
  if(MyDestinationIsNear()) slowDown();
  if(MyDestinationIsHere()) goInLowerLane();
else //myLane==2 or more
  if(ICanGoInLowerLane())
    if(GoingInLowerLaneIsForcedOrConvenient())
      goInLowerLane();
    else // I cannot go in lower lane
      if(MyDestinationIsNear())
        slowDown(); goInLowerLane();
      if(!HaveSpaceProblemsForward()) //every lane
        if(TakeoverIsPossibleAndMyDestinationIsFar())
          if(TakeOverIsDesired()) takeover();
          else followTheQueue();
        else followTheQueue();
      else // I have space problems forward
        if(TheTakeoverIsForced()) takeover();
  return;
END;
```

References

1. S. Di Gregorio and R. Serra, *Fgcs* **16**, 259 (1999).
2. A. Schadschneider, *Phys. A* **372**, 142 (2006).
3. K. Nagel and M. Schreckenberg, *J. Phys. I* **2**, p. 2221 (1992).
4. D. E. Wolf, *Phys. A* **263**, 438 (1999).
5. W. Knospe, L. Santen, A. Schadschneider and M. Schreckenberg, *J. Phys. A: Math. Gen.* **33**, 477 (2000).
6. M. Lárraga, J. del Ríob and A. lcaza L., *Transport. Res. C.* **13**, 63 (2005).
7. S. Di Gregorio and D. C. Festa, *Cellular Automata for Freeway Traffic*, in *Proceedings of the First International Conference Applied Modelling and Simulation*, ed. G. Mesnard 1981, pp. 133–136.
8. S. Di Gregorio, D. Festa, R. Rongo, W. Spataro, G. Spezzano and D. Talia, *A microscopic freeway traffic simulator on a highly parallel system*, Advances in Parallel Computing Vol. 11 1996, pp. 69–76.

IMAGINARY OR ACTUAL ARTIFICIAL WORLDS USING A NEW TOOL IN THE ABM PERSPECTIVE

PIETRO TERNA

Department of Economic and financial sciences, University of Torino,
Corso Unione Sovietica 218bis, 10134 Torino, Italy
terna@econ.unito.it

We propose SLAPP, or Swarm-Like Agent Protocol in Python, as a simplified application of the original Swarm protocol, choosing Python as a simultaneously simple and complete object-oriented framework. With SLAPP we develop two test models in the Agent-Based Models (ABM) perspective, building both an artificial world related to an imaginary situation with stylized chameleons and an artificial world related to the actual important issue of interbank payment and liquidity.

1 From a "classical" protocol to a new tool

1.1 The Swarm protocol

The Swarm protocol dates from the mid-1990s, so in some way it is a "classical" reference in the relatively young world of the agent-based simulation, mainly for social sciences. The Swarm project (www.swarm.org), born at Santa Fe Institute, has been developed with an emphasis on three key points [1]: (i) Swarm defines a structure for simulations, a framework within which models are built; (ii) the core commitment is to a discrete-event simulation of multiple agents using an object-oriented representation; (iii) to these basic choices Swarm adds the concept of the "swarm," a collection of agents with a schedule of activity.

The "swarm" proposal was the main innovation coming from the Swarm project, diffused as a library of function together with a protocol to use them. Building the (iii) item required a significant effort and time in code development by the Swarm team; now using Python we can attain the same result quite easily and quickly.

To approach the SWARM protocol via a clear and rigorous presentation it is possible refer to the original SimpleBug tutorial [2], developed using the Objective-C programming tool (built on C and Smalltalk, www.smalltalk.org) by

Chris Langton and the Swarm development team; the tutorial also has explanatory texts in the README files of the main folder and of the internal subfolders). The same contents have also been adapted by Staelin [3], to the Java version of Swarm, and by myself [4], to create a Python implementation, exclusively related to the protocol and not to the libraries. Note that the Swarm original libraries are less important, anyway, using a modern object-oriented language. The SWARM protocol can be considered as a meta-*lingua franca* to be used in agent-based simulation models.

1.2 The Swarm-Like Agent Protocol in Python (SLAPP)

The SLAPP project [a] has the goal of offering to scholars interested in agent-based models a set of programming examples that can be easily understood in all its details and adapted to other applications.

Why Python? Quoting from its main web page: "Python is a dynamic object-oriented programming language that can be used for many kinds of software development. It offers strong support for integration with other languages and tools, comes with extensive standard libraries, and can be learned in a few days."

Python can be valuably used to build models with a transparent code; Python does not hide parts, have "magic" features nor have obscure libraries. Finally, we want to use the openness of Python to: (i) connect it to the R statistical system (R is at http://cran.r-project.org/; Python is connected to R via the rpy library, at http://rpy.sourceforge.net/); (ii) go from OpenOffice (Calc, Writer, …) to Python and vice versa (via the Python-UNO bridge, incorporated in OOo); (iii) do symbolic calculations in Python (via http://code.google.com/p/sympy/); and (iv) use Social Network Analysis from Python, with tools like the Igraph library (http://cneurocvs.rmki.kfki.hu/igraph/), the libsna library (http://www.libsna.org/), and the pySNA code (http://www.menslibera.com.tr/pysna/).

The main characteristics of the code reproducing the Swarm protocol in Python are introduced step by step via the on line tutorial referenced above. Summarizing:

- SimpleBug - We use a unique agent run the time by the way of a *for* cycle, without object-oriented programming.

---

[a] Python is at www.python.org. You can obtain the SLAPP tutorial files and the related examples at: http://eco83.econ.unito.it/terna/slapp.

- SimpleClassBug - We run the time by the way of a *for* cycle, now with object-oriented programming to create and to use a unique agent as an instance of a class; in this way, it is quite simple to add other agents as instances of the same class.

- SimpleClassManyBugs - We run the time by the way of a *for* cycle, with object-oriented programming to create many agents as instances of a class; all the agents are included in a collection; we can interact with the collection as a whole.

- SimpleSwarmModelBugs - We run the time following a *schedule* of events based on a simulated clock; the schedule can be managed in a dynamic way (events can change sequences of events). With object-oriented programming we create families of agents as instances of classes, within a special class, the model class, that can be considered as a the experimental layer of our program.

- SimpleSwarmObserverBugs - As above, but we now have the model and all its items (agents, schedule, clock) included in a top layer of the application, that we name "observer". The observer runs the model and uses a schedule to apply tools like graphical representations, report generations, etc. The clock of the observer can be different from that of the model, which allow us to watch the simulation results with a granularity different from that of the simulated events.

In the object-oriented programming perspective the starting module generates the observer as an instance of the observer class. The observer creates: (i) the reporting tools, (ii) one or more models as instances of the class model and (iii) finally the schedule coordinating the top-level actions. Each model creates (iv) the instances of the agents and (v) the schedule controlling their behavior.

2 Creating an imaginary artificial world: the surprising behavior of learning chameleons

2.1 The structure of the chameleon world

A test model created with SLAPP involves intelligent chameleons. We have chameleons of three colors: red, green and blue. When two chameleons of different colors meet, they both change their color, assuming the third one. If all chameleons change to be the same color, we have a steady-state situation. This case is possible, although rare. Even so, what if the chameleons of a given color want to conserve their identity? On the other hand, what if they strongly want to change it?

With the on-line version of the chameleon model, [b] we can see the chameleons moving randomly in their space. The chameleons can (i) move in a random way or (ii) refer to a *runner* mind (nine neural networks able to evaluate the moves from the point of view of the runners, used together) to avoid changing their color, or (iii) refer to a *chaser* mind (nine neural networks able to evaluate the moves from the point of view of the chasers) to look for contacts and so to change their color, if they want to change their color. As an example, if we tell a specific type of chameleons (i.e., the red ones) to be conservative, adopting the rules generated by a reinforcement learning process to avoid contacts, they become capable of increasing in number with the strategy of decreasing their movement when staying in zones free from chameleons of other colors, and getting close to subjects with their own color.

2.2 Reinforcement learning

We use reinforcement learning [5] to develop intelligent behavior in our chameleons. Rewards and punishments come from the experience made in the past by chameleons while acting.

The evaluation of the rewards is quite simple. We consider here only a 3×3 space, as in Figure 1, with nine potential moves, accounting also for staying in the initial patch. The red chameleon in the central position of the internal 3×3 square in Figure 2 has three enemies around it; moving to the north in the center of the green square, it would have only two (visible) enemies, with a +1 reward; the reward can also be negative. Here we always consider two steps jointly and sum up their rewards, without the use of a discount rate. The double-step analysis compensates for the highly bounded rationality strategy applied both to the knowledge of the world (limited in each step to a 5×5 space) and the evaluation of the rewards, in a 3×3 area.

---

[b] The project, built in SLAPP, has been implemented also in NetLogo, relatively to the on-line visual representation of the results: with NetLogo you can easily produce an applet to be directly run in a browser; NetLogo is at http://ccl.northwestern.edu/netlogo/.
The model is at http://eco83.econ.unito.it/terna/chameleons/chameleons.html, where you can find also an animation with voice explanations.
I thank Riccardo Taormina, a former student of mine, for developing this kind of application with great involvement and creativity. Many thanks also to Marco Lamieri, a former Ph.D. student of mine, for introducing the powerful chameleon idea.

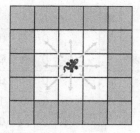

Figure 1. The nine possible moves of the agent (staying in place and moving to one of the eight adjacent squares).

We have nine neural networks, one for each of the potential moves in the 3×3 space, with 25 input elements (the 5×5 space), ten hidden elements and a unique output, which is a guess of the reward that can be gained by choosing each one of the nine valid moves in the presence of each specific situation in the 5×5 space. Neural networks are used to summarize the results of the described trial and error process, avoiding the requirement of dealing with huge tables reporting the data upon the reward of each potential choice in each potential situation.

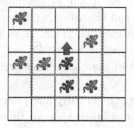

Figure 2. Two subsequent spatial situations, always considered from the center of the two internal 3x3 squares.

The dimension of the observed space, 5x5, obeys the bounded rationality principle. We could also consider a 7x7 space, but the number of potential situations increases enormously: in the case of the 5x5 space, omitting the central square, we have 2^{24} potential situations. With a 7x7 space, we have 2^{48} possibilities, a number on the order of 3 times 10^{14}. In the 5x5 case, we are able to explore a simplified space like that of Figure 3, with about 17 million of possible states, with symmetric inputs not mandatory to produce symmetric outcomes (the decision system attributed to our agents is intrinsically imperfect).

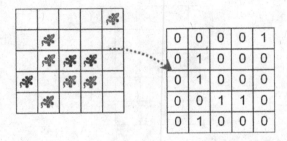

Figure 3. The chameleon in the center of the grid (a red one) creates the matrix corresponding to its observed world. Values of 1 identify the places of the chameleons of colors other than red.

2.3 Results in chameleon behavior

The chameleons can (i) move in a random way, or (ii) refer to a *runner* mind (the nine neural networks able to evaluate the moves from the point of view of the runners, used together) both to avoid contacts and to avoid changing their color, or (iii) refer to a *chaser* mind (the nine neural networks able to evaluate the moves from the point of view of the chasers, used together) to look for contacts and to change their color.

The running model can be found at the address reported in the note introducing the section, with and animation with voice instruction to train users in interacting with the simulation interface. It is possible to reproduce the different behavior of both the running and of the chasing chameleons, remembering that the actions of the chameleons are arising from an automatic learning process.

The model can also be metaphorically interpreted in the following way: an agent diffusing innovation (or political ideas or financial innovative behaviors) can change itself through interaction with other agents. As an example, think about an academic scholar working in a completely isolated context or, on the contrary, interacting with other scholars or with private entrepreneurs to apply the results of her work. On the opposite side, an agent diffusing epidemics modifies the others without changing itself.

3 Recreating an actual world in an artificial way: interbank payments

3.1 The payment problem

From the imaginary world of the chameleons, always using SLAPP, we shift to focus on the concrete aspects of an actual banking system, recreating the interaction of two institutions (a payment system and a market for short-term liquidity) to investigate interest rate dynamics in the presence of delays in interbank movements. [c]

The problem is a crucial one because delays in payments can generate liquidity shortages that, in the presence of unexpected negative operational or financial shocks, can produce huge domino effects [6]. Here, we use agent-based simulation as a magnifying glass to understand reality.

3.2 Two parallel systems

We have two parallel and highly connected institutions: the RTGS (Real Time Gross Settlement payment system) and the eMID (electronic Market of Interbank Deposit). Considering the flow of interbank payments settled via the first institution, we simulate delays in payments and examine the emergent interest rate dynamics in the eMID. In this kind of market the interest rate is the price. A few microstructures of the market should be investigated and understood.

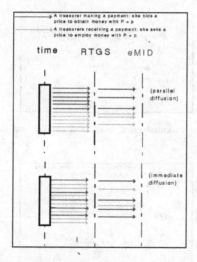

Figure 4. Events to RTGS and from RTGS to eMid, in two different ways: a "quasi UML" representation of a sequence diagram.

[c] I am deeply grateful to Claudia Biancotti and Leandro D'Aurizio, of the Economic and Financial Statistics Department of the Bank of Italy, and to Luca Arciero and Claudio Impenna, of the Payment System Oversight Office of the Bank of Italy, for having involved me in considering this important problem. The model is a contribution that I hope can be useful in identifying some aspects of the problem in a complexity perspective. You can obtain the code of the model by e-mailing the author.

In Figure 4 we present a modified representation of the standard sequence diagram of the UML (Unified Modeling Language, www.uml.org) formalism, introducing time as the first actor or user in the sequence. Events come from a time schedule to our simulated environment; the treasurers of the banks, acting via the RTGS system, with given probabilities bid prices, to buy liquidity in the eMID, or ask prices, to sell liquidity in the same market. Bid and ask probabilities can be different. The simple mechanism of bidding or asking on a probabilistic basis (if and only if a payment has to be done or has been received, as in Figure 4), will be integrated – in future developments - with an evaluation of the balance of the movements in a given time period.

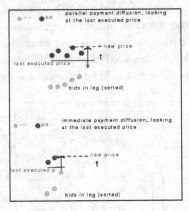

Figure 5. Looking at the last executed price, both in a parallel and in an immediate diffusion scheme.

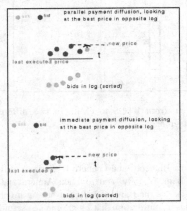

Figure 6. Looking at the best proposal in the opposite market log, both in a parallel and in an immediate diffusion scheme.

The different sequences of the events (with their parallel or immediate diffusion, as in Figure 4) generate different lists of proposals into the double-action market we are studying. Proposals are reported in logs: the log of the bid proposals, according to decreasing prices (first listed: bid with the highest price); and the log of the ask proposals, according to increasing prices (first listed: ask with the lowest price). "Parallel" means that we are considering an actual situation in which all the treasurers are making the same choice at practically the same time.

In Figure 5 we discuss how a new price is proposed to the market when we look at the last executed price as a reference point, placing a price below it to get a ask position easily matched. In this case, both the case of parallel proposal and that of immediate diffusion are figuring out close expected situations. On the other hand: (i) this is not the case in the logs of the unmatched proposals, with ask prices greater than bid prices; (ii) the behavior of a market maker, not present here, is based on positive ask minus bid price differences. Other potential microstructures have to be investigated.

In Figure 6, a new price is proposed to the market looking at the best proposal in the opposite log as a reference point, placing a price below it to get an ask position easily matched. The cases of parallel proposal and that of immediate diffusion are now producing quite different effects.

3.3   A case of simulated behavior

We show here an interesting case of the dynamics emerging from this simulation environment, that occurs when the diffusion of the payment into the RTGS system is parallel and the operators look at the last executed price in the eMID. The run reported in Figure 7 shows a non-trivial behavior of the interest rate. The dynamic is here magnified due to the dimension chosen for micro-movement in bids and asks. In these five days, we have a huge movement of this time series, as a consequence of significant delays in interbank payments. The simulation runs step by step, but we account for breaks in the time to reproduce the end of each day (i.e., cleaning all the positions, etc.).

Elaborating the interest rate series with the standard AR (autoregressive) and MA (moving average) technique, directly connecting SLAPP to R as seen above, we find in the graphs of the second row in Figure 8 a typical AR(1) model. On the contrary, in a situation like that of Figure 9, with data coming from a simulation run in which no payment delays occur, we find a random-walk dynamic in the interest rate first differences (first graph of the second row), without any correlation evidence.

254

This analysis suggests that the strong difference between these two situations is the direct consequence of the presence/absence of the payment delays.

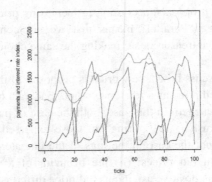

Figure 7. Interest price dynamic (upper line), stock of due payments (intermediate line) and flow of the received payments (lower line) in case of relevant delay in payments (with a uniform random distribution between 0 and the 90% of the available time until the time break). Time breaks at 20, 40, ... (end of a day).

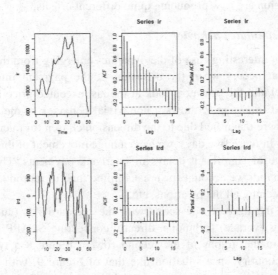

Figure 8. The autocorrelation analysis of the interest rate data of Figure 7 (with the presence of delays in interbank payments). First row: raw data; lagged correlations among data; the same, but as partial correlations. Second row: data first differences with lag 1; their lagged correlations; their lagged partial correlations.

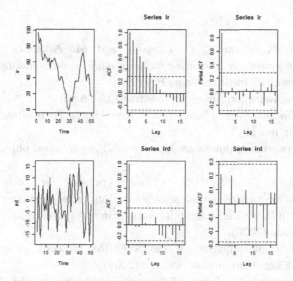

Figure 9. The autocorrelation analysis of the interest rate data in a case of absence of delays in interbank payments). First row: raw data; lagged correlations among data; the same, but as partial correlations. Second row: data first differences with lag 1; their lagged correlations; their lagged partial correlations.

4   Future developments

There are three promising lines for future developments:

- In terms of SLAPP, the development of the capability of directly probing each agent, the graphical representation of spatial dynamics and of social networks links, and the simplification of the schedule code for event dynamic.

- In terms of chameleons, the introduction of communication capabilities, mainly via information left in the space, to search for the emergence of social coordination and of social capital.

- In terms of the payment system, applying also in this case the reinforcement learning technique, the introduction of a *market maker*, i.e., a subject continuously asking and bidding a price for selling and buying money, with a spread, assuring liquidity to the market and developing a pricing capability aimed at the containment of liquidity crisis.

References

1. N. Minar, R. Burkhart, C. Langton, and M. Askenazi, *The Swarm simulation system: A toolkit for building multi-agent simulations.* WP 96-06-042, Santa Fe Institute, Santa Fe (1996), http://www.swarm.org/images/b/bb/MinarEtAl96.pdf.

2. C. Langton and Swarm development team, Santa Fe Institute, *SimpleBug tutorial*, on line at http://ftp.swarm.org/pub/swarm/apps/objc/sdg/swarmapps-objc-2.2-3.tar.gz, (1996?).

3. C. J. Staelin, *jSIMPLEBUG, a Swarm tutorial for Java*, (2000), at http://www.cse.nd.edu/courses/cse498j/www/Resources/jsimplebug11.pdf, only text, or http://eco83.econ.unito.it/swarm/materiale/jtutorial/JavaTutorial.zip, text and code (2000).

4. P. Terna, implementation of the SWARM protocol using Python, at http://eco83.econ.unito.it/terna/slapp, (2007).

5. R. S. Sutton, A. G. Barto, *Reinforcement Learning: An Introduction.* Cambridge MA, MIT Press (1998).

6. L. Arciero, C. Biancotti, L. D'Aurizio and C. Impenna, Exploring agent-based methods for the analysis of payment systems: a crisis model for StarLogo TNG. *Journal of Artificial Societies and Social Simulation*, forthcoming, (2008).

PART VI

SYSTEMS BIOLOGY AND SYNTHETIC BIOLOGY

HOW CRITICAL RANDOM BOOLEAN NETWORKS MAY BE AFFECTED BY THE INTERACTION WITH OTHERS

C. DAMIANI, A. GRAUDENZI, M. VILLANI

Department of Social, Cognitive and Quantitative Sciences,
Modena and Reggio Emilia University, 2100 Reggio Emilia, Italy
E-mail: {chiara.damiani, alex.graudenzi, marco.villani}@unimore.it
www.cei.unimore.it

In previous articles we have introduced Multi Random Boolean Networks (MRBNs) as a possible model for the interaction among cells within multicellular organisms or within bacteria colonies. MRBNs are sets of Random Boolean Networks (RBNs), placed on a Cellular Automaton, whose nodes may be affected by the activation of some nodes in neighbouring networks. In this paper we study the effects induced by interaction on the dynamics of those RBNs that - if isolated - lay in the critical region. It is shown that the influence of interaction is not univocal; nevertheless it is possible to identify three classes of representative behaviours. RBNs belonging to each class seem to have different dynamical peculiarities even in isolation: although sharing the parameters proper of critical networks, they substantially differ in their typical response to perturbations.

Keywords: Genetic network model; Random Boolean network; Cellular automaton; Interaction; Cell-criticality.

1. Introduction

Random Boolean Networks have been introduced in the 70s, by Stuart A. Kauffman,[1] as a model for gene regulatory networks. Despite their extreme simplicity, RBNs became a paradigm in complex system biology[2] and more generally in the study of complex systems. Since there are many excellent reviews on RBNs, in Section 1 we only introduce a few key definitions and properties.

The most intriguing feature of the Random Boolean Network model is its effectiveness in addressing the issue of criticality. The idea that living beings are in a critical state has been proposed as a general principle by different authors and with different meanings.[3-6] The description of the dynamics of Random Boolean Networks allows precise statements about this hypothesis to be made, by providing an accurate, although not all-embracing, definition of cell criticality. Furthermore, the quantitative information obtainable from the statistical analysis of collection of RBNs allows to test the hypothesis of cell-criticality against experimental data, which are nowadays largely available thanks to new technologies such as DNA microarrays. At this regard, it has recently been shown that Random Boolean networks can accurately describe the statistical properties of perturbations in the expression of the whole set of genes of S. cerevisiae after single knock-outs.[7-9] Since the distribution of avalanches in gene expression data turned out to be related to a parameter which also determines the network dynamical regime, the results allowed to give new support to the hypothesis that cells are in an ordered regime close to the critical boundary. The critical regime is plausibly the one which contributed to the evolution of cells, allowing those transformations that may have been favoured by natural selection; indeed the features of this regime make it neither so stable to neutralize every mutation nor too unstable to make a perturbation dramatically spread.

Arguments in favour of cell criticality apply to isolated cells or to organisms as a whole;[10,11] on the other hand unicellular organisms tend to form colonies where all cells grow close to each other and communicate by transferring some molecules. Intercellular communication is even more intense in multicellular organisms where cells interact in order to form higher level structures like tissues.

It is therefore interesting to study the relationship between the dynamics of an isolated cell and that of a collection of interacting cells. If we consider a single RBN as a model for an isolated cell, than we can investigate the issue by studying how the dynamics of a RBN is affected by the interaction with other RBNs. The Multi Random Boolean Networks model has this purpose. A MRBN is a 2-D cellular automaton each of whose site hosts a complete RBN. Each RBN can interact with the networks in a given neighbourhood. The interaction among RBNs is inspired to biological cellular communication. Although cells communicate through several and highly sophisticated mechanisms, in this preliminary study we concentrate our attention on a rough approximation of the particular mechanism according to which adjacent cells share some proteins able to cross the cellular membranes. The communication is thus modelled by letting the activation of some nodes be affected by the expression of "corresponding" nodes (e.g. nodes coding for the same proteins) in neighbouring cells. Section 3 will describe the model in detail.

A real cell expresses only a small subset of its genes and the existence of the different cell types that we find in multicellular organisms depends on the differentiation in gene expression patterns. Patterns of genes (nodes) expression can be found also in RBNs and they correspond to the attractors of a network. It is therefore sensible to study the interaction among structurally identical RBNs and analyze their possibly different attractors. We concentrated, *in primis*, in finding under which conditions a given set of interacting cells can be found in the same attractor (while the analysis provides also information about conditions which allow the coexistence of different attractors, no attempt is made here to analyze the spatial patterns which may appear).

Preliminary investigations[12–14] have showed that the dynamics of MRBNs is far from trivial. While one might have guessed that interaction might either enhance or reduce the likelihood of finding cells in the same attractor, the simulation results showed that different behaviours are possible for different RBNs, notwithstanding their structural similarity (all analyzed RBNs are created according to the parameters typical of the critical regime). It has been possible to recognize three classes of representative behaviours. Section 4 gives a short description of these classes.

This paper aims at analyzing the dynamics of the isolated critical RBNs belonging to the three different classes in order to disentangle possible individual peculiarities that might be enhanced by interaction. The approach is that of studying their response to perturbations. Section 5 shows how the results of the study suggest a relationship between a RBN response to perturbations and its response to interaction.

2. Random Boolean Networks

There are some excellent reviews on RBNs;[10,15,16] in this section we will only outline the main features of classical RBNs, which are utilized in our MRBN model.

A RBN is an oriented graph composed by N Boolean nodes. Every node corresponds to a gene of a specific genetic network, and it may be active ($value = 1$) or inactive ($value = 0$). The influences among genes are represented by directed links. In a classical RBN each node has the same number of incoming connections k_{in}, chosen at random with uniform probability among the remaining $N - 1$ nodes. It then turns out that the distribution of outgoing connections per node follows a Poisson distribution.

The activation value of a certain node depends on the value of its input nodes, according a specific Boolean function. Each Boolean function is generated at random, assigning to any combination of its inputs the output value 1 with probability p (also called the *bias*). Both the topology and the Boolean function associated to each gene do not change in time; the network dynamics is discrete and synchronous.

In order to analyze the properties of an ensemble of random Boolean networks, different networks are synthesized and their dynamical properties are examined. While individual realizations may differ markedly from the average properties of a given class of networks,[17] one of the major discoveries is the existence of two different dynamical regimes, an ordered and a disordered one, divided by a "critical zone" in parameter space. The dynamical regime of a RBN depends primarily on two parameters, the average connectivity of the network $< k_{in} >$ and the bias p. For a more detailed discussion, the reader is referred to.[10,17–19]

3. Multi Random Boolean Networks

Many interesting cell aggregates, including epithelium, bacterial colonies and in-vitro cell cultures, have an essentially two-dimensional structure. We can therefore model the interaction among RBNs on a two-dimensional space. A Multi Random Boolean Network is a 2D square lattice cellular automaton[20] with M^2 cells, each of them being occupied by a complete Random Boolean Network. The neighbourhood is of the von Neumann type (composed by the cell itself and its N, E, S, W neighbours). We assume wrap around so the overall topology is toroidal. In order to model the interaction among cells with the same genetic material, we consider a copy of the same RBN in each site of the automaton. Therefore, every RBN of the automaton is structurally identical, while the initial activation states may differ. In particular, all the RBNs have:

(1) the same number (N) of Boolean nodes;
(2) the same topology, i.e. the same ingoing and outgoing connections for each node of the network;
(3) the same Boolean functions associated to each node.

Hence, these common features define the *genome* of a RBN. Since all the RBNs are genetically identical we well refer to it also as the *the genome of a MRBN*. Given the structural identity of the cells, we can define corresponding nodes in different cells in a straightforward way: suppose e.g. that node i gets inputs from node 40 and 80 in cell A, and that it sends its output to nodes 1 and 99. Then in every other cell node i will have incoming links from 40 and 80 and outgoing links to 1 and 99 as well.

The fact that some molecules can pass from one cell to another is modelled here by assuming that the state of some genes in a cell can be affected

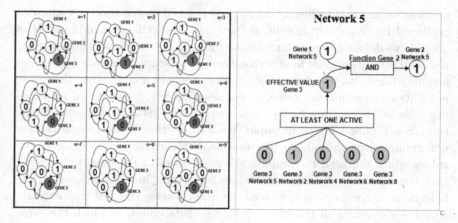

Fig. 1. Example of a 3*3 automaton. Networks in the automaton are identified by a number from 1 to 9. Gene 3 (grey coloured) is chosen to be a communication node. On the right we can see a section of network number 5 (central one): the input values for gene 2 are the elementary value of gene 1 and the effective value of gene 3 (gene 3 owns elementary value = 0 and effective value = 1). The effective value of gene 3 in the network 5 is 1 because gene 3 of the neighbouring network number 2 is active.

by the activation of some other genes of a neighbouring cell.[a] Those nodes whose products can cross the membrane, and therefore influence neighbouring cells, will be called *communication nodes* (or genes). They are a subset of the total number of nodes of the RBN. Since all the RBNs are structurally identical, the subset of communication nodes is exactly the same for all the cells of the automaton. We define the *elementary value* of every node to be the value computed according its Boolean function and the values of its input nodes, belonging to the same RBN. Thus, the elementary value does not take into account intercellular communication. On the contrary, the *effective* value of a communication node is calculated taking into account also what happens in neighbouring cells: suppose that gene i is controlled by genes 40 ad 80, and that it is a communication node. Then the elementary value of gene i in cell J is computed from the values of genes 40 and 80 in cell J, but its effective value depends also upon the values of genes 40 and 80 in the cells which are close to J. We define the *interaction strength* between neighbouring cells as being equal to the fraction f of com-

---

[a]In other works[14] we model cell-to-cell interaction taking inspiration from an alternative mechanism according to which a molecule can bind to a membrane receptor, thus triggering a cascade of events in the cell which hosts the receptor.

munication nodes. The idea is that a gene in a cell senses also the molecules produced by neighbouring cells, if they cross the cellular membrane. It is possible to define different interaction rules. In this work we suppose that if a communication gene in is active, then its product is present also in the neighbouring cells. Therefore, the effective value of a communication node is 1 if its own elementary value, or the elementary value of the corresponding node of at least one of its neighbouring cells, is 1 (*at-least-one-active* rule). See Figure 1 for an example. We also define the effective value of a non communication node to be equal to its elementary value. Hence, the values which are used to compute the next states of the network coincide with the effective values. It is therefore important, in order to analyze the behaviour of the networks, to consider the effective state of the network, which is the vector of the effective values of its nodes. In much the same way, the elementary state of the network is defined as the vector of the elementary values of its nodes. Note that, if we consider a single cell at time t, its elementary state at time $t + 1$ is determined by its effective state at time t.

4. Dynamics of MRBNs

As described in the previous section the structure of a MRBN is defined by:

- The dimension of the lattice M;
- The genome G of each RBN (given by the number of nodes N, the topology and the Boolean functions);
- The interaction strength R.

It is interesting to study how the dynamics of MRBNs is affected by the variation of the strength of the interaction. In order to isolate the effects of interaction it's convenient to keep the other structural elements fixed, while varying the interaction strength. The number of different attractors reached by the networks of a MRBN, for different levels of f, is particularly worth of note. Such an analysis is complicated by the fact that in several experiments some cells do not reach an attractor within the limitations which have been imposed in the simulations. [b] We therefore found it useful

---

[b]The search for an attractor starts after 200 time steps (a step being a complete update of each node of each RBN present in the automaton) and the maximum possible period (for attractor search) is set to 200: in this way, we do not consider attractors whose period is higher than 200 steps. The search ends when an attractor is found or when the system reaches 2000 steps (in this case we state that the RBN has found no attractor).

to introduce a variable defined as follows:

$$\Omega \equiv N_A + N_W \tag{1}$$

where N_A is the number of different attractors present in the automaton at the end of the experiment and N_W is the number of cells of the automaton which do not reach any attractor at the end of the experiment. Their sum can heuristically be considered as a measure of how disordered a MRBN is: Ω is a variable which takes its minimum value (1) when the MRBN is maximally ordered (all the cells reach the same attractor) and which takes its maximum value (equal to the total number of cells M) when the automaton is disordered (all the cells do not reach any attractors or reach different attractors).

The analysis of the dependency of Ω on the interaction strength for a large sets of MRBNs with genomes typical of critical networks [c] has revealed unexpected and intriguing dynamics. The sample considered is a set of 900 20*20 MRBNs, all composed of RBNs of 100 nodes each one. Each MRBN evolves with different levels of the interaction strength (11 levels ranging from 0 to 1, step 0.1).[d] We observed a variance of Ω raising with the interaction strength, suggesting that different MRBNs might have different response to interaction. An in-depth analysis has established the existence of different dependencies of Ω on the interaction strength which can, nevertheless, be grouped in three representative behaviours (see Figure 2):

- *Constant MRBNs*: Ω is constant and equal to 1: all the networks of the MRBN reach the same attractor, independently of the value of f and also in absence of interaction. 15% of the observed MRBNs lay in this class and the attractors of their RBNS are fixed points;
- *Growing MRBNs*: Ω increases as f increases (18% of the sample);
- *Bell-shaped MRBNs*: in most cases (56% of the sample) the dependency of Ω on f is bell-shaped, with a single maximum for $f \notin \{0, 1\}$.

---

[c]The networks are classical RBNs, with an equal number of incoming connections per node ($k_{in} = K = 2$). The input nodes are chosen at random with uniform probability among the remaining $N - 1$ nodes, auto and multiple connections being forbidden. The Boolean functions are generated at random independently for every node, with no restrictions and $bias = 0.5$. The initial states of the nodes are chosen at random for every RBN, independently from those of the other cells

[d]Every simulation on every MRBN is repeated 150 times. For other details on the realization of the experiments see.[13]

The remaining 11% of the MRBNs of the sample do not clearly belong to one of the previous groups and are labelled as not classifiable (NC).

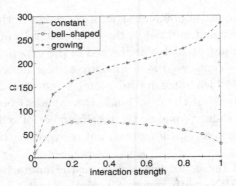

Fig. 2. Dependency upon f of average Ω for the 3 different classes of MRBNs. Note that the dependency for *constant MRBNs* is the straight line $\Omega = 1$.

The three groups of MRBNs have been studied also in relation to other order indicators (e. g. the fraction of runs where all the cells of the automaton reach the same attractor; the fraction of runs where all the cells reach an attractor; the fraction of runs where no cell reaches an attractor). The dependency of these aggregate variables on f seems to be consistent with Ω.[13]

5. Relationship between response to interaction and response to perturbations

The different behaviours that RBNs may show when they interact through MRBNs induced us to investigate possible explanations of such differences. We decided to study the dynamical features of the isolated RBNs which have lead to the *constant, bell-shaped* and *growing MRBNs* studied in the previous section; for convenience we will refer to the three classes of isolated RBNs respectively as *constant RBNs, bell-shaped RBNs* and *growing RBNs*. The intent was to find some peculiarities which may be representative of the RBNs belonging to each class. One possible method to investigate the dynamical regime of complex systems in general is to analyse their response to perturbation. A large sensitivity to perturbations is usually related to disordered (or chaotic) systems, while, vice versa, a low sensitiveness (higher robustness) refers to ordered systems. A perturbation may be caused by the *flip* of one node chosen at random, i.e. the change of the activation value

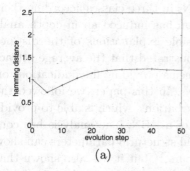 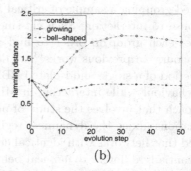

Fig. 3. (a) Variation of average Hamming distance in time for the whole sample of RBNs. (b) Variation of average Hamming distance in time for the 3 different classes of RBNs.

of one node in the initial condition of the network. A measure of the sensibility to perturbations can be borrowed from the information theory: the *Hamming distance (HD)*, which is the number of differing bits between two strings. A possible means to discriminate the dynamical regimes of RBNs is to observe the variation in time of the Hamming distance between a control network (wild type WT) and a perturbed network (PN). A Hamming distance tending to 0 is peculiar of ordered networks; a value close to 1 is related to critical networks, while values higher than 1 refers to chaotic ones.[18] Figure 3-b reports the average value of the hamming distance in time (after a flip) for the three classes of RBNs. [e] Although the average variation of the Hamming distance on the whole set of networks (Figure 3-a) closely resembles the results of past studies of the same kind,[18] it is possible to observe well distinct behaviour for the three classes. In particular:

- *Constant RBNs* tend to neutralize perturbations in time (HD tend to 0);
- *Bell-shaped RBNs* tend to maintain the initial perturbation without spreading it (HD keeps close to the initial value 1);
- In *growing RBNs* the number of nodes affected by a flip grows in time till reaching a maximum level after a transient.

6. Conclusions

The identification of three distinct representative behaviours for critical RBNs, when interacting through Multi Random Boolean Networks, has

---

[e]Note that each node of the network may be flipped, included *communication nodes*.

allowed grouping a sample of critical RBNs in three classes according their response to interaction. Such classification has induced an in-depth analysis of each group in order to find possible explanations of the distinct behaviours. In previous works[13] we demonstrated that the average attractor period of a single and isolated RBN seems providing fine indications on its behaviour while interacting in MRBNs. In this paper, we followed the approach that involves the study of perturbations, which is able to provide indications about the dynamical regime of a RBN. The analysis has confirmed that networks with identical critical structural parameters can show substantially different dynamical behaviours,[17] but it has also shown that these behaviours are similar for RBNs within each of the three individuated classes and well-distinct among RBNs belonging to different classes. In particular: the representative trend of the diffusion of perturbations for those RBNs which have been defined *constant RBNs* closely resembles that of ordered networks; the one of *bell-shaped RBNs* is rather similar to that of critical networks, while *growing RBNs* seem to behave like chaotic networks. A conceivable suggestion is a possible analogy between perturbations in gene expressions and intercellular signalling among RBNs. Indeed, interaction with neighbouring cells (according to the *at-least-one-active* rule) is likely to turn an inactive communication node on; this might remind a flip. Thus, communication may be advantageous only for those RBNs effectively critic (which would correspond to *bell-shaped RBNs*). This is a preliminary hypothesis and needs a deeper examination.

A direction for further research is the study of the similarity of the attractors reached by the cells of a MRBN, through a cluster analysis. It would also be interesting to examine which are the structural peculiarities of RBNs whose dynamics is effectively critical. Further extensions of the model may include a mechanism for the duplication of cells, an essential phenomenon that is strictly related to intercellular communication.

Acknowledgments

This work has been partially supported by the Italian MIUR-FISR project nr. 2982/Ric (Mitica).

References

1. S. A. Kauffman, *Top. Dev. Biol.* **6**, 145 (1971).
2. K. Kaneko, *Life: An Introduction to Complex Systems Biology* (Springer, 2006).

3. C. G. Langton, Computation at the edge of chaos: Phase transitions and emergent computation, in *Emergent Computation*, ed. S. Forrest (MIT Press, Cambridge, MA, 1991)
4. P. Bak, C. Tang and K. Wiesenfeld, *Physical Review A* **38** (1988).
5. P. Bak, *How Nature works* (Springer, New York, 1996).
6. S. A. Kauffman, *At home in the universe* (Oxford University Press, 1995).
7. R. Serra, M. Villani and A. Semeria, *Journal of Theoretical Biology* **227**, 149 (2004).
8. R. Serra, M. Villani, A. Graudenzi and S. A. Kauffman, *Journal of Theoretical Biology* (2007).
9. P. Ramo, J. Kesseli and O. Yli-Harja, *Journal of Theoretical Biology* (2006).
10. S. A. Kauffman, *The origins of order* (Oxford University Press, 1993).
11. S. A. Kauffman, *Investigations* (Oxford University Press, 2000).
12. M. Villani, R. Serra, P. Ingrami and S. Kauffman, Coupled random boolean network forming an artificial tissue, in *Cellular Automata*, , Lecture Notes in Computer Science Vol. 4173/2006 (Springer Berlin / Heidelberg, 2006).
13. R. Serra, M. Villani, C. Damiani, A. Graudenzi, A. Colacci and S. A. Kauffman, Interacting random boolean networks, in *Proceedings of ECCS07: European Conference on Complex Systems*, eds. J. Jost and D. Helbing 2007
14. R. Serra, M. Villani, C. Damiani, A. Graudenzi and A. Colacci, The diffusion of perturbations in a model of coupled random boolean networks, in *ACRI 2008*, ed. H. Umeo (Springer Lecture Notes in Computer Science, Berlin, 2008). In press.
15. I. Harvey and T. Bossomaier, Time out of joint: Attractors in asynchronous random boolean networks, in *Proceedings of the Fourth European Conference on Artificial Life (ECAL97)*, eds. P. Husbands and I. Harvey (MIT Press, 1997).
16. M. Aldana, S. Coppersmith and L. Kadanoff, Boolean dynamics with random couplings, in *Perspectives and Problems in Nonlinear Science (Springer Applied Mathematical Sciences Series)*, eds. E. Kaplan, J. E. Marsden and K. R. Sreenivasan 2003
17. U. Bastolla and G. Parisi, *Physica* **D**, 45 (2003).
18. M. Aldana, *Physica* **D**, 45 (2003).
19. J. E. S. Socolar and S. A. Kauffman, *Phys Rev. Lett.* **90** (2003).
20. S. E. Yacoubi, S. Chopard and S. Bandini (eds.), *Cellular Automata: 7th Int. conf. on Cellular Automata for research and Industry*, LNCS Vol. 4173, 2006.

DYNAMICS OF INTERCONNECTED BOOLEAN NETWORKS WITH SCALE-FREE TOPOLOGY

C. DAMIANI and M. VILLANI

Department of Social, Cognitive and Quantitative Sciences
Modena and Reggio Emilia University
2100 Reggio Emilia, Italy
E-mail: {chiara.damiani, marco.villani}@unimore.it
www.cei.unimore.it

CH. DARABOS and M. TOMASSINI

Information Systems Institute ISI - HEC, University of Lausanne,
CH-1015 Lausanne, Switzerland
E-mail: {christian.darabos, marco.tomassini}@unil.ch
www.hec.unil.ch/isi

In this paper we investigate how the dynamics of a set of coupled Random Boolean Netowrks is affected by the changes in their topology. The Multi Random Boolean Networks (MRBN) is a model for the interaction among Random Boolean Networks (RBN). A single RBN may be regarded as an abstraction of gene regulatory networks, thus MRBNs might represent collections of communicating cells e.g. in tissues or in bacteria colonies. Past studies have shown how the dynamics of classical RBNs in the critical regime is affected by such an interaction. Here we compare the behaviour of RBNs with random topology to that of RBNs with scale-free topology for different dynamical regimes.

Keywords: Artificial Tissue; Random Boolean Networks; Scale-Free Networks; Biological Networks Dynamics

1. Introduction

A major challenge of biology is to unravel the organization and interactions of cellular networks that enable complex processes such as the biochemistry of growth or cell division. Molecular biology, based upon a reductionist approach, has been crucial in understanding individual cellular components and their functions, but it does not offer adequate concepts and methods to comprehend how the interaction among the components rise to the function and behaviour of the entire system. A system-wide perspective on component interactions is thus required as a complement to the classical approach. Hence, one of the goals of Complex Systems Biology is to discover new emergent properties that may arise from the systemic view in order to better understand the entirety of processes that happen in a biological system.[1]

One of modern biology's main tasks is to detect, understand and model the topological and dynamical properties of the network, formed by the several components that interact with each other through pairwise interactions, which control the behaviour of the cell. At a higher level of abstraction, the components can be reduced to a series of nodes that are connected to each other by links, with each link representing the interactions between two components. The nodes and links together form a network or, in a more formal mathematical language, a graph.

An interesting dynamical network model is the Random Boolean Networks (RBN), proposed by Kauffaman[2] over 30 years ago as a model for gene regulatory networks. Despite its simplifying assumptions the model has contributed to lead to the realization that, notwithstanding the importance of individual molecules, cellular function is a contextual attribute of strict and quantifiable patterns of interactions between the myriad of cellular constituents. The very rich and unexpected behaviour found for the model is the existence of a dynamical phase transition controlled by two parameters, the average connectivity K and the probability p that a node is active (the *bias*).

Although RBNs model qualitatively points in the right direction, it fails to account for a quantitative description of what has been recently observed in genetic networks. One of the main problems is that the critical connectivity is very small for most values of the bias. It has however been shown that the fine-tuning usually required to achieve stability in Boolean Networks with a totally random topology is no longer necessary when the network topology is scale-free.[3]

There are several excellent reviews on the impact of a scale-free topology on the dynamics of Random Boolean Networks.[3-5] In this paper we face a different issue: how the dynamics of scale-free RBNs is affected by the interaction with other networks. The topic ought to be central to Complex Systems Biology, since cells rarely act as isolated organisms and the regulatory network of a cell accounts, therefore, also for the expression of genes in other surrounding cells, which may be cells of the same multi-cellular organism as well as other unicellular organisms living in the same colony. In order to simulate this phenomena we make use of a model introduced by some of us: the Multi Random Boolean Networks (MRBNs).[6,7]

A MRBN is a cellular automaton where each site hosts a complete RBN. The interaction among the cells of a given neighbourhood is modelled by letting the activation of some nodes in a cell depend on that of nodes in surrounding networks according specific interaction rules. Preliminary analysis involving classical RBNs has shown that the dynamics of the model is rather intricate. RBNs with similar structural features may behave in extremely different ways. Some indicator of order suggest that in some cases a RBN can become less ordered while interacting with others.[6,7]

The analysis described in this preliminary work approaches the investigation of whether the dynamical peculiarities of RBNs with scale-free topologies, their robustness *in primis*, have implications also in the way they are affected by interaction with others networks through a MRBN. Results will then be compared to those of *classical* RBNs. In the next section, we describe RBNs as proposed by Kauffmann, then, in section 3 we describe *classical* RBNs with scale-free structure, that tries to account for recent findings in real-life networks. In section 4 we will describe the Multi Random Boolean Network model that will be studied by statistical sampling using numerical simulation in section 5. Finally, in section 6 we give our conclusions

2. Introduction to Random Boolean Networks

In a RBN with N nodes, a gene is represented as a binary node, meaning that it is expressed if it is on (1), and it is not otherwise (0). Each gene receives inputs from K randomly chosen other genes. This involves a perfectly homogenous degrees distribution where all the nodes are connected exactly to K nodes. Initially, one of the possible Boolean functions of K inputs is assigned at random to each gene. The network dynamics is discrete and synchronous: at each time step all nodes simultaneously examine their inputs, evaluate their Boolean functions, and find themselves in their new states at the next time step. The RBN's state is the states of all its node at a time t. The system travels in time through its phase space, until a point or cyclic attractor is reached. Then it will either remain in that point attractor forever, or it will cycle through the states of the periodic attractor. Since the system is finite and deterministic, this will happen at most after 2^N time steps. An RBN attractor is therefore an ensemble of network states that loop back, the number of states constituting an attractor is called the attractor's cycle length.

Here we summarize the main results of the original model.[8,9] First of all, as some parameters are varied such as K, or the probability p of expressing a gene, the RBN can go through a phase transition. Indeed, for every value of the connectivity K, there is a critical value of the probability p such that the system is in the ordered regime, in the chaotic regime or on the edge of chaos. Kauffman found that for $K = 2$ and $p = 0.5$, the number of different attractors of the networks scales as the square root of N and the mean cycle length scales at most linearly with N for $K = 2$. Kauffman's conjecture is that attractors correspond to cell types in the RBN phase space, and only those attractors that are short and stable under perturbations would be of biological interest. Thus, according to Kauffman, $K = 2$ RBNs laying at the edge between the ordered and the chaotic phases can be seen as abstract models of genetic regulatory networks. In the ordered regime, the period length is found to be independent of N and usually short (around 1) and in the chaotic regime, the attractors length grows exponentially. Neither of these regimes show biological plausibility. We believe that the original views of Kauffman are still valid, provided that the model is updated to take into account present knowledge about the structure of real gene regulatory networks without loosing its attractive simplicity.

3. Scale-Free Boolean Networks

RBNs are directed random networks. The edges have an orientation because they represent a chemical influence from one gene onto another, and the graphs are random because any node is as likely to be connected to any other node in an independent manner. There are two main types of RBNs, one in which the connections are random but the degree K is fixed and a more general one in which only the average connectivity $\bar{k}$ is fixed. Random graphs with fixed connectivity degree were a logical generic choice in the beginning, since the exact couplings in actual genetic regulatory networks were unknown. Today it is more open to criticism since it does not correspond to what we know about the topology of biological networks. According to present data, as far as the *output* degree distribution is concerned [a] many biological networks seem to be of the scale-free type or of a hierarchical type[11-13] but not random. The *input* degree distributions seem to be close to normal or exponential instead. A scale-free distribution for the degree means that $p(k)$ is a power law $P(k) \sim k^{-\gamma}$, with γ usually

---

[a] The *degree distribution function* $p(k)$ of a graph represents the probability that a randomly chosen node has degree k.[10] For directed graphs, there are two distributions, one for the outgoing edges $p_{out}(k)$ and another for the incoming edges $p_{in}(k)$.

but not always between 2 and 3. In contrast, random graphs have a Poisson degree distribution $p(k) \simeq \bar{k}^k e^{-\bar{k}}/k!$, where $\bar{k}$ is the mean degree, or a delta distribution as in a classical fixed-degree RBN. Thus the low fixed connectivity suggested by Kauffman ($K \sim 2$) for candidate stable systems is not found in such degree-heterogeneous networks, where a wide connectivity range is observed instead. The consequences for the dynamics may be important, since in scale-free graphs there are many nodes with low degree and a low, but not vanishing, number of highly connected nodes.[10,14]

Aldana presented a detailed analysis of a Boolean network model with scale-free topology.[3] He has been able to define a phase space diagram for scale-free Boolean networks, including the phase transition from ordered to chaotic dynamics, as a function of the power law exponent γ. He also made exhaustive simulations for several relatively small values of the network size N. In our model we have thus adopted networks with a scale-free output distribution, and a Poisson input distribution, as this seems to be closer to the actual topologies. In section 3.1 we shall give details on the construction of suitable graphs of this type for our simulations.

3.1. *Construction of Scale-Free Networks*

As said above, Kauffman's RBN are directed graphs. Let's suppose that each node i ($i \in \{1, \ldots, N\}$) receives k_i^{in} inputs and projects a link to k_i^{out} other nodes, i.e. there are k_i^{out} nodes in the graph that receive an input from node i. Among the N nodes of the graph, the distribution $p_{in}(k)$ of the input connections is not necessarily the same of the distribution of the output connections $p_{out}(k)$. In fact, and as anticipated in the preceding section, according to present data many biological networks, including genetic regulatory networks, suggest a not homogeneous output distribution and a Poisson or exponential input distribution.[11–13] Whether $p_{in}(k)$ is Poisson or exponential is almost immaterial for both distributions have a tail that decays quickly, although the Poisson distribution does so even faster than the exponential, and thus both have a clear scale for the degree. On the other hand, $p_{out}(k)$ is very different, with a fat tail to the right, meaning that there are some nodes in the network that influence many other nodes (hubs). We will be using networks that follow a scale-free output distribution, where $p(k)$ is the probability a node n will have a degree k:

$$p(k) = \frac{1}{Z} k^{-\gamma}$$

where the normalization constant $Z(\gamma) = \sum_{k=1}^{k_{max}} k^{-\gamma}$ coincides with Riemann's Zeta function for $k_{max} \to \infty$. The output distribution approximates a normal function centered around $\bar{k}$. Figure 1 offers a taste of what such distributions look like.

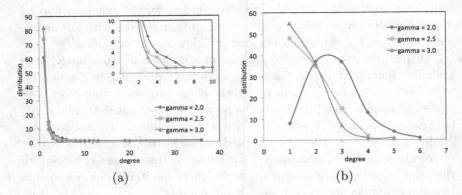

Fig. 1. Actual output degrees distributions (a) and input degrees distributions (b) of a sample generated networks of size $N = 100$ and $\gamma = 2.0$, $\gamma = 2.5$ and, $\gamma = 3.0$. Distributions are discrete and finite; the continuous lines are just a guide for the eye. In (b), the influence of γ on the Poisson distribution is due to its influence on the mean degree $\bar{k}$ of the network.

4. The Multi Random Boolean Networks model

A Multi Random Boolean Network is a two-dimensional square cellular automaton with M^2 cells, each of them being occupied by a complete Random Boolean Network. Cells in multi-cellular organisms or in bacteria colonies share the same genetic material, hence in order to model the interaction among cells in such structures we have to consider the interaction of systems that are structurally identical. Thus all the RBNs of the automaton have:

- the same number N of Boolean nodes;
- the same topology, i.e. the same ingoing and outgoing connections for each node of the network;
- the same Boolean functions associated to each node.

They may only differ as for their initial configuration (state of each node at time $t = 0$). Each RBN can interact with its N, E, S, W neighbours. The spatial shape of the automaton is a torus: cells at extreme sides interact with those at the opposite end.

The interaction among neighbouring RBNs is inspired by biological cellular communication. Biological cells communicate trough several and highly sophisticated procedures, it would be therefore possible to define a variety of interaction mechanisms. In this specific study we consider a rough approximation of the particular mechanism according to which adjacent cells share some proteins able to cross the cellular membranes.[b].

Given the identical structure of the cells, we can define "corresponding" nodes in different cells as nodes coding for the same proteins. The communication is modelled by letting the activation of some nodes be affected by the expression of "corresponding" nodes in neighbouring cells. Those nodes whose products can cross the membrane, and therefore influence neighbouring cells, will be called *communication nodes* (or genes). Note that the subset of communication nodes is exactly the same for all the cells of the automaton. We call *interaction strength* (IS) the fraction of communication nodes on the total number of nodes for each network.

As described above, the value of a node is computed according to its Boolean function using the value of its input nodes. We call this the *elementary value*. When we take into account intercellular communication, we define the *effective value* of a communication node only, inputs from RBNs in neighbouring cells according to a specific interaction rule as well as the that of input nodes of the same RBN. In this work we suppose that a communication node has effective value 1 if its own elementary value is 1 or at least one of the corresponding nodes in neighbouring cells. The idea is that a gene in a cell senses also the molecules produced by neighbouring cells, when they cross the cellular membrane. We define the effective value of a non-communication node to be equal to its elementary value. Hence, the values used to compute the next states of the network coincide with the effective values.

For sake of simplicity, let us name *MRBN-RND*, Multi Random Boolean Networks containing Random topology RBNs and *MRBN-SF*, Multi Random Boolean Networks containing Scale-Free topology RBNs.

---

[b]In other works[7] we model cell-to-cell interaction taking inspiration from an alternative mechanism according to which a molecule can bind to a membrane receptor, thus triggering a cascade of events in the cell which hosts the receptor.

5. Methodology and Simulations

In order to explore the space of both MRBNs types in their different regimes, we have produced a sample of 50 networks of 100 nodes each in every possible combination of topologies (scale-free or random) and regimes (order, edge of chaos, and chaos, when possible). Each network was then replicated on a 20×20 MRBN. In the case of random networks, we have used Kauffmann[2] assumption that the phase transition between regimes is governed by the value of the so-called average expected sensitivity $S = 2p(1-p)\bar{k}$, where $\bar{k}$ is the average network connectivity and p the bias. For $S < 1$ the network is in the ordered regime and for $S > 1$ it is in the chaotic regime. In our case $K = \bar{k} = 2$, thus values of $p_{order} \neq 0.5$ and $p_{edge\_of\_chaos} = 0.5$. As for scale-free networks, we have used the following values of $\gamma_{order} = 3.0$, $\gamma_{edge\_of\_chaos} = 2.5$ and $\gamma_{chaos} = 2.0$ in the degree distributions, as suggested by Aldana.[3] In total, we have 5 groups of 50 different networks, each of which is placed in a MRBN as described above in section 4. Then, for each level of IS $\in [0.0, 1.0]$ by steps of 0.1, we repeat the experiment on 150 different sets of initial configurations. We record the following values at the end of each run:

- the fraction of simulations α where all the cells have reached the same attractor;
- the fraction of simulations β where all the cells have reached an attractor;
- the fraction of simulations δ where none of the cells has reached an attractor.

We will also analyze whether the change in the parameters discriminating a regime of an isolated RBN (the bias p for classical RBNs and the scale-free exponent γ for scale-free topologies) affect in any way the dynamics of the system as a whole. For sake of fluency we will use in an interchangeably the terms *regime* of a RBN and *regime* of MRBN, although the regime of a MRBN may not coincide with that of RBNs in isolation.

5.1. *Analysis of the Results*

In this first study, we will compare and contrast the behavior of MRBN-SF against that of MRBN-RND in terms of α, β and δ as defined in the previous section. This will give us an insight on how easily RBNs converge to an attractor under different ISs. Moreover, we will be able to see how well all the individual cells of a MRBN can "synchronize" in stabilizing by which we mean finding the same attractor. Figure 2 shows the results of computer simulations of all groups of MRBNs. On each graph we show α, β and δ. The left-hand side column corresponds to MRBN-SFs and the right-hand side one represents MRBN-RBNs. Each line shows results for different regimes of the interacting RBNS, from top to bottom: order, edge of chaos and chaos (in case of $K_{out} = 2$ the chaotic regime does not exists for MRBN-RBNs).

When comparing the different regimes for each categories of MRBN, we notice that the regime seems to have a much greater influence on scale-free topologies than on random ones. In fact, the differences in behavior of MRBN-RNDs is moderate, and confirm the results[6] that increasing the IS the system seems to get more ordered (α rises), whereas we note a drop in the number of systems where all cells reach any attractor. Deeper analysis have revealed that this apparently paradoxical phenomena is due to the fact that the dynamical behaviour of each single realization may be highly different from the average dynamics of RBNs[15] and increasing the IS seem to further intensify these individual peculiarities. Nevertheless, the observation that β is higher than α, which is, in turn, higher than δ is true for all regimes.

On the other hand, MRBN-SFs have very different dynamic behaviours depending on the regime they evolve in. MRBN-SFs that are in the ordered regime show the same tendency of α to increase with IS, typical of MRBN-RDNs, although in this case values are higher and span from 0.4 to 0.6. Systems where no cell reaches a stable attractor (δ curve) drop almost to zero, and systems where all the cells reach an attractor (β) decrease as the IS grows.

For the MRBN-SFs closer to the so-called *edge of chaos*, we observe a drop in beta more important than the decrease in the random topology systems. Moreover, the cases where no cell in the system reaches an attractor are much more frequent than in the MRBN-RNDs, and are systematically higher than the synchronisation cases (α curve). On the other hand, we witness a significant decrease in the presence of systems synchronizing to the same attractor (α curve) as the IS increases, as in the MRBN- RNDs, although the values remain very low (between 0.1 and 0.2).

In the chaotic regime MRBN-SFs systems struggle to relax to any attractors (δ above 0.9) as soon as interaction is added, ever so slightly. Interestingly, it seems that in the few cases where all cells of a system reach an attractor, they actually seem to all reach the same one (α and β coincide). But the biggest difference among MRBN-SFs and MRBN-RNDs is surely

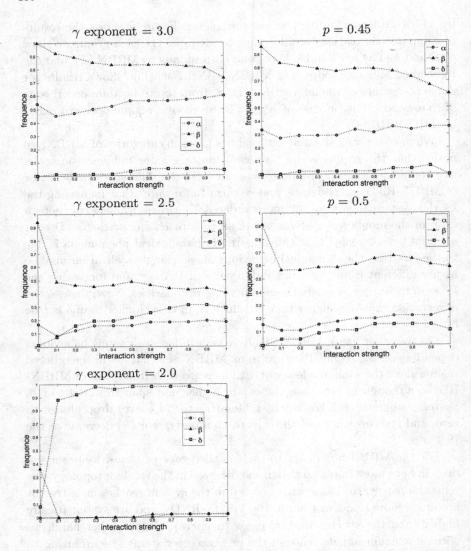

Fig. 2. Comparing results over 50 networks and 150 runs of scale-free topologie RBNs (left-handside column) and random topologie RBNs (right-handside column). Each line corresponds to a different regime, from top to bottom: order, edge of chaos and chaos. Note that with $K_{out} = 2$ the latter case is presented only by MRBN-SFs.

the impossibility of MRBN-RNDs to obtain chaos in case of $K_{out} = 2$. In any case, preliminary simulations of both systems with $K_{out} = 4$ seem to indicate that the fraction of cases where no cell reaches an attractor is always significantly smaller in MRBN-RNSs than in MRBN-SFs systems.

A general observation is that, compared to MRBN-RNDs systems, MRBN-SFs seem to have a higher level of order in ordered regimes (high synchronisation level - α curve), but are significantly more prone to chaos when the fraction of hubs increases (edge-of-chaos and chaotic regimes). The unique cases in which order is found (i.e. attractors found) are single state attractors, called fixed point attractors. As a result, the α and β curves coincide in Figure 2 where $\gamma = 2.0$ (chaotic regime).

A possible explanation involves the high transmission capabilities of hubs, the presence of which may significantly:

- influence the probability of involving communication nodes;
- when regarding communication nodes, increase the areas involved in changes coming from other cells.

As a conclusive remark, we observe that both systems show analogous behaviours with respect to the increase of IS.

6. Conclusions

The analysis of a model of interconnected Random Boolean Networks has shown how the interaction deeply affects networks' dynamics. We have compared the impact of interaction for RBNs with random and with scale-free topology.

This work is only a first step toward the characterisation of the influence of topologies different from that of Erdos-Renyi on the behaviour of coupled genetic systems. Nevertheless, we show that not homogeneous connectivity distributions of the genetic regulatory network of single cells have a significant influence on the behaviour of these systems seen as a whole, a point that, if verified in further works, can lead to the individuation of a possible situation sensible to the influence of evolutionary processes. In fact, under the hypothesis that the previous simulations capture some key aspects of the reality, it is possible that the step from unicellular beings toward multicellular ones is not costless, also with respect to the topological strategies used on the genetic control processes by the different living species. Of course, further work is required, involving both different topologies and different complex coupling interactions.

Acknowledgements

M. Tomassini and Ch. Darabos gratefully acknowledge financial support of the Swiss National Science Foundation under contract 200021-107419/1.

References

1. K. Kaneko, *Life: An Introduction to Complex Systems Biology* (Springer, 2006).
2. S. A. Kauffman, *J. Theor. Biol.* **22**, 437 (1969).
3. M. Aldana, *Physica D* **185**, 45 (2003).
4. C. Oosawa and M. A. Savageau, *Physica D* **170**, 143 (2002).
5. R. Serra, M. Villani and L. Agostini, *Physica A* **339**, 665 (2004).
6. R. Serra, M. Villani, C. Damiani, A. Graudenzi, A. Colacci and S. A. Kauffman, Interacting random boolean networks, in *Proceedings of ECCS07: European Conference on Complex Systems*, ed. D. H. J. Jost, (CD-Rom, paper 35) 2007
7. R. Serra, M. Villani, C. Damiani, A. Graudenzi and A. Colacci, The diffusion of perturbations in a model of coupled random boolean networks, in *ACRI 2008*, ed. H. Umeo (Springer Lecture Notes in Computer Science (in press), Berlin, 2008).
8. S. A. Kauffman, *The Origins of Order* (Oxford University Press, New York, 1993).
9. M. Aldana, S. Coppersmith and L. P. Kadanoff, Boolean dynamics with random couplings, in *Perspectives and Problems in Nonlinear Science*, eds. E. Kaplan, J. E. Marsden and K. R. SreenivasanSpringer Applied Mathematical Sciences Series (Springer, Berlin, 2003).
10. M. E. J. Newman, *SIAM Review* **45**, 167 (2003).
11. A. Vázquez, R. Dobrin, D. Sergi, J.-P. Eckmann, Z. N. Oltvai and A.-L. Barabàsi, *Proc. Natl. Acad. Sci USA* **101**, 17940 (2004).
12. R. Albert, *J. of Cell Science* **118**, 4947 (2005).
13. C. Christensen, A. Gupta, C. D. Maranas and R. Albert, *Physica A* **373**, 796 (2007).
14. R. Albert and A.-L. Barabasi, *Reviews of Modern Physics* **74**, 47 (2002).
15. U. Bastolla and G. Parisi, *Physica* **D**, 45 (2003).

A NEW MODEL OF GENETIC NETWORK:
THE GENE PROTEIN BOOLEAN NETWORK

ALEX GRAUDENZI, ROBERTO SERRA

Department of Social, Cognitive and Quantitative Sciences,
University of Modena and Reggio Emilia,
viale allegri 9 Reggio Emilia, I-42100, Italy

A new model of genetic network, named Gene Protein Booelan Newtork (GPBN), is introduced and described.

The model is a generalization of the well-known random Boolean network model (RBN) and has been implemented in order to allow comparisons with real biological data coming from DNA microarray time course experiments.

In the classical RBN model the timing of biochemical mechanisms is not taken into account. On the contrary, in the GPBN model the link among every gene and its correspondent protein is made explicit and every protein is characterized by specific synthesis and decay times, that influence the overall dynamics. Preliminary simulations on different sets of critical networks seem to suggest that the larger is the memory of the network, the more ordered the system would tend to be.

1. Introduction

Simple models of gene regulatory networks have repeatedly proven fruitful in the description of the emerging dynamics that such deeply complex systems show, as well as in the detection of intriguing and often not expectable generic properties.

One of the most relevant examples is the well-known random Boolean networks (RBN) model, introduced by Stuart A. Kauffman in the late 60's.[1] Despite a clearly unrealistic simplicity of the model itself, the spectacular advances in the molecular biology techniques of the last decades recently allowed to attest its effectiveness in describing some of the typical dynamical behaviours of real networks,[2-6] moreover confirming some of the visionary hypotheses which had been put forth by Kauffman many decades ago, only counting on the results of limited simulations.

One important example is the description of the response of genetic networks to perturbations, which can be nowadays operated and measured on real genomes by means of the technique of DNA microarray [e.g.[7]]. The simplest possible model of genetic network indeed provides surprisingly accurate match on the distribution of the "avalanches" of (knockout) perturbation, besides allowing to get a significant experimental support to the intriguing hypothesis of living beings evolving in a dynamical regime close to the "edge of chaos".[4,5]

Nevertheless, and in spite of an always growing amount of experimental data coming from molecular biology, the scarcity of the *right* data to compare with such a model represented, so far, a limit for its application and for the possibility of drawing definitive conclusions.

The model which will be introduced in this paper has been thought in order to allow comparison with a broader set of experimental data (in particular those which refer to so-called time courses, where the expression levels of thousands of genes are measured at close time intervals) without however losing the clear and effective approach of the original Kauffman model.

In detail, the model aims to fill one of the limitations of the RBN model, that is the absence of the timing of the molecular processes at the basis of the mechanism of gene regulation. In real cells, proteins need time to be synthesized indeed and every protein is characterized by a specific decay time. In the classical RBN model the synthesis of proteins is considered to be instantaneous and their decay time is the same, as it is implicit in the synchronous update rule. On the contrary, in the model we are introducing here, the link among every gene and its protein is made explicit and every single protein is characterized by specific synthesis and decay times.

The details of the model will be presented in section 2. In section 3 the features of the simulations will be introduced and in section 4 the preliminary results will be shown. Remarks and conclusions are in section 5.

2. The model

For an exhaustive description of the classical model of RBN, the reader is referred to.[2,8,9] Here we will present the main features of the new model, named *Gene Protein Boolean Network* (*GPBN*) model.

A Gene Protein Boolean Network is an oriented graph with two distinct types of entities: gene nodes and protein nodes. Both elements are present in the same number, N, and both elements are of the Boolean kind (there are N gene - protein pairs). If the value of a gene node is 1 (0) this means that the gene is active (inactive), while if the value of a protein node is 1 (0) this means that the protein is present (absent) in the cell.

There is one and only one outgoing link from each gene node to a protein node. If the value of a gene node is 1 (gene active) at time t, the value of the correspondent protein will be 1 (i.e. the protein will be present in the cell) at time step $t+1$. In practice, we assume the protein synthesis time as the measure unit for the time steps. We suppose, on the contrary, that the presence or the absence of certain proteins inside the cell instantaneously influences the activation of the genes of the network (i.e. at the same time step). Notice that the indices of gene nodes and protein nodes are strictly related: gene node i is linked to protein node i (in gene - protein pair i).

In gene regulatory networks, a particular protein can act as an activator or an inhibitor, alone or in combination with other proteins, for a specific gene. In the model this is represented by outgoing links from protein nodes to gene nodes. Every protein node owns a number of outgoing links ranging from 0 to N (multiple connections are forbidden, while links to the gene which synthesizes the protein itself are allowed[a]).

The value of a specific gene node at time t is determined on the basis of the value of its input protein nodes at time t, according a specific Boolean function which is associated to that gene node, and which is chosen at random, according to some probability distribution; the simplest choice is that of a uniform distribution among all the $2^{2^{k_{in}}}$ possible Boolean functions of k_{in} arguments.

Moreover, every protein node is characterized by a specific *decay time*, defined as follows. Once a protein node is activated at time t through the activation of its correspondent gene node at time $t-1$, the decay time is the number of time steps in which the value of the protein node remains equal to 1, notwithstanding the possible inactivation of its gene node. Hence, a protein node is characterized by a specific *decay phase* at each step of the dynamics, the decay phase being an integer number ranging from 0 to the precise decay time of that protein node. The decay phase decreases of one unit at time t if the related gene node at time $t-1$ is inactive and becomes equal to the decay time if the gene node at time $t-1$ is active. If the decay phase of a protein node at time t is larger than 0, then the value of the protein node is 1. Notice, so, that we assume the decay time of the proteins to be a multiple of the unitary synthesis time.

The topology of the network, the Boolean function associated to each gene node and the decay time associated to the protein nodes do not change in time. The network dynamics is discrete and the update is synchronous.

---

[a]In classical RBNs auto-connections are forbidden. We removed this constraint on the basis of recent biological studies that describe as plausible the auto-regulation of a gene (alone or in combination with other genes).

Let $g_i(t) \in \{0,1\}$, $p_i(t) \in \{0,1\}$ and $d_i(t) \in \{0, dt_i\}$, $i = 1,2,...N$ denote respectively the activation values of the gene node, protein node and decay phase of the gene-protein pair i at time t, (being dt_i the specific decay time of protein node i) and let $S(t) = [g_1(t), g_2(t)...g_N(t); p_1(t), p_2(t)...p_N(t); d_1(t), d_2(t)...d_N(t)]$ represent the whole state of the network at time t.

The system is deterministic and, as in the case of classical RBNs, the attractors are cycles of finite length. The average connectivity is defined as the ratio between the number of connections from protein nodes to gene nodes and the number of gene - protein pairs.

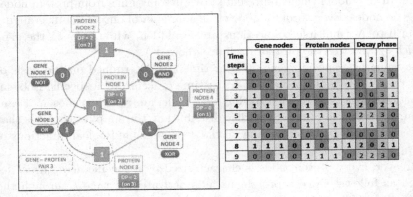

Time steps	Gene nodes 1	2	3	4	Protein nodes 1	2	3	4	Decay phase 1	2	3	4
1	0	0	1	1	0	1	1	0	0	2	2	0
2	0	0	1	1	0	1	1	1	0	1	3	1
3	1	0	0	1	0	0	1	1	0	0	3	1
4	1	1	1	0	1	1	1	0	2	0	2	1
5	0	0	1	0	1	1	1	0	2	2	3	0
6	0	0	1	0	1	1	1	0	1	1	3	0
7	1	0	0	1	0	0	1	0	0	0	3	0
8	1	1	1	0	1	0	1	1	2	0	2	1
9	0	0	1	0	1	1	1	0	2	2	3	0

Fig. 1. *(left)*Scheme of an example GPBN. Circle nodes represent gene nodes (Boolean value), while squared nodes represent protein nodes *(idem)*.*(right)* The dynamics of the example GPBN of the left figure for the first 9 time steps (4 gene - protein pairs). The state of the system is composed by three distinct vectors, representing respectively the Boolean activation values of the gene nodes, those of the protein nodes and the decay phase corresponding to the protein nodes. Gene i is related to protein i, and the decay phase i belongs to protein i. The link among gene nodes and protein nodes, the Boolean functions, the decay times and the initial condition of the nodes and the decay phases are those of the example network in the left figure. In this example the network reaches its limit cycle after a transient = 4, since the states of the network in step 4 and step 8 (in bold) are identical. Therefore, the limit cycle is 4 steps long.

The dynamics of the systems can be schematized as follows.

- Time 0 (initial condition)
  - A random Boolean value is associated to each protein node of the network.
  - If the value of the protein i is 1, a random value of the correspondent decay phase is given, ranging from 1 to the specific time decay. Otherwise the value of the decay phase is 0.

- The Boolean value of every node at time 0 is determined on the basis of the values of its input protein nodes at time 0, according its specific Boolean function.

- Time t, $t > 0$; for every i:
    - If the gene node i is active at time $t - 1$, the specific time decay is assigned to the decay phase of the protein node i, otherwise the value of the decay phase at time t is decreased of 1 unit.
    - If the gene node i is active at time $t - 1$ OR the value of the phase decay i at time t is > 0, the value of the protein node i at time t becomes 1, otherwise it will be 0.
    - The Boolean value of every gene node at time t is determined on the basis of the values of its inputs protein nodes at time t, according its specific Boolean function.

One of the main differences with the classical RBN model lays in the features of the limit cycles of the system. Since the whole state of the network is given by three vectors (those of gene nodes, protein nodes and decay phases), the dynamic of the vector of the gene nodes only is not deterministic (nor is that of the vector of protein nodes).

Let us note that, like in the case of classical RBNs, networks with the same structural parameters can show deeply different dynamical behaviours, often very different from the average one. Therefore, the relevance of analyses on single networks is rather limited[10] and it is necessary to perform a statistical analysis of a large number of simulation runs over a large number of networks with the same structural parameters. In this way one can study the generic properties of such systems and the effects of the variation of the specific parameters on the dynamics.

Note that the choice of a decay time equal to 1 for all the protein nodes leads to a configuration of the GPBNs which is that of classical random Boolean networks. Therefore, GPBN model can be regarded as a generalization of RBN model.

3. The simulations

Preliminary simulations, whose results are summarized below, regarded networks with the following structural parameters and features.

All the networks are composed by $N = 100$ gene-protein pairs; the ingoing connections are equal to 2 for each gene node; on the other hand the outgoing connections from the protein nodes are randomly distributed on all the genes with uniform probability (thus, the outgoing connections distribution follows a Poisson distribution[2]). The Boolean function associated to each node is chosen with uniform probability among all the 16 Boolean functions. These parameters were chosen for two reasons. First, because an ingoing connectivity fixed to 2 and a set of Boolean functions with no restriction are the standard parameters of the classical RBN model.[2] Second, because these parameters are those of classical RBNs characterized by a critical behaviour.[9] Therefore, we considered interesting to primarily investigate the dynamical behaviour of GPBNs with these fixed parameters and different values of the maximum decay time.

In detail, the decay time is randomly associated to each protein node with uniform probability in a range that goes from 1 to an arbitrary integer value defined as *maximum time decay* (MDT). Networks with a maximum decay time = 1, 2 and 4 have been simulated. We built $M = 40$ GPBNs with the features specified above and for each of them we varied the value of MDT while keeeping fixed the other features. In this way we could isolate the influence of the variation of the maximum decay time on the dynamics, avoiding the possibile noise due to the change of the other parameters.

4. The results

The variables we analyzed in order to understand the influence of a change in the choice of the maximum decay time (MDT) on the overall dynamics are the number of different attractors reached by the networks (over a certain number of simulation runs characterized by different initial conditions), the average period of the attractors and the average length of the transients (i.e. the number of time steps before reaching the attractor).

We will first take a look at the average values of these critical variables for the different sets of networks, according to different values of MDT.

Looking at table 1 we can notice how the average number of different attractors reached by the networks substantially decreases as the maximum decay time increases. The trend is less defined when analyzing the variation of the average period of the attractors as well as the length of the transients in the three different cases.

To gain further insight, we analyzed the distributions of the differences of the three variables of table 1 in the different kinds of networks (fig. 2). Looking at the distribution of the differences in the number of different attractors, for both the comparisons ($MDT = 2$ vs. $MDT = 1$ and $MDT = 4$ vs. $MDT = 2$), even if the large majority of networks decrease the number of attractors or has no changes at all (coherently with the average value

Table 1. Variation of the average number of attractors, of the average period of the attractors and of the average transient (over the total number of simulated networks), for sets of networks characterized by different values of the maximum decay time (MDT), i.e. 1, 2 and 4.

MDT	Av. numb. of attractors	Av. attractors' length	Av. transient
1	8.0	19	31
2	3.8	15	24
4	2.1	16	27

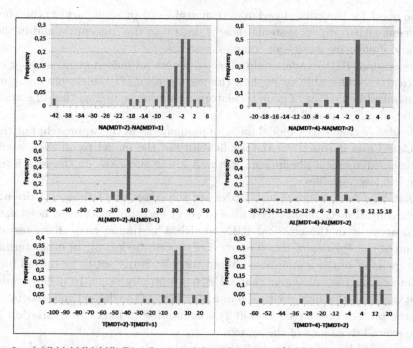

Fig. 2. (a)(b)(c)(d)(e)(f). Distribution of the differences of [(a)(b) the number of attractors (NA); (c)(d)the average length of the attractors (AL); (e)(f) the average transient (T)] calculated on each specific simulated network with different values of MDT [(a)(c)(e) NA, AL and T of networks with MDT = 2 - NA, AL and T of the same networks with MDT = 1; (b)(d)(f) NA, AL and T of networks with MDT = 4 - NA, AL and T of the same networks with MDT = 2], over the total number of different simulated networks.

seen in table 1), some of the networks actually show an increase in the number of their attractors, as well as some other ones show smaller values very far from the average one. Also for what concerns the length of the periods and of the transients, ahead of the general trend given by the average values given in table 1, one can observe variations even deeply far from the medium one, either larger or smaller.

Once again, networks with the same structural parameters can show indeed different dynamical behaviours. Nevertheless, these first results allowed us to hypothesize that higher values of MDT would lead on the average to more orderer behaviours, with a smaller number of different attractors.

5. Conclusions

The quest for the critical variables responsible for (a large part of) the phenomena that complex systems show is central in the conception and in the study of effective models of complex systems.

The original RBN model could allow to detect an indeed small number of critical variables able to explain (at least a part of) the deeply complex dynamical behaviour of gene regulatory networks, as well as that of a wide range of different real systems.

The idea of the GPBN model finds its fundaments in this approach, but the declared aim is to extend the applicability of the model to an even wider range of dynamical phenomena, without losing the focus on the search for the few key variables.

What is actually central in GPBNs are the concepts of time and timing. Biochemical mechanisms need time to occur and this aspect was not taken into account in the classical RBN model. The timing that characterizes some of the entities of the systems (i.e. protein nodes) or, in other terms, the memory of the system itself about its past, turns out to be indeed fundamental in the behaviour that GPBNs eventually show. In other words, the variation of a single critical parameter such as the decay time associated to the protein nodes (see section 2) accounts for an important variation in the overall dynamics of the system.

As specified in section 4, not all the networks behave in the same way in response to the variation of the range of the decay time of the protein nodes: once more the heterogeneity of networks with the same structural parameters is confirmed.

Nevertheless, these preliminary analyses on networks characterized by critical parameters seem to suggest that the larger is the memory of the network (i.e. the maximum decay time), the more ordered the system would tend to be. The consequences of this possible conclusion clearly involve the intriguing issue of the criticality of the networks, with a particular reference to the biological plausibility and suitability of the model.

However, and before drawing definitive conclusions, further researches and analyses are needed to better study and understand this new model. Besides exploring the intrinsic features of the model, particular interest will be devoted to the comparison with experimental data coming from microarray time course experiments, that could allow to test the real effectiveness of the model in describing real networks.

Aknowledgments

This work has been partially supported by the Italian MIUR-FISR project nr. 2982/Ric (Mitica).

References

1. Kauffman, S.A. *Gene regulation networks: A theory of their global structure and behaviour.*, Top. Dev. Biol. 6, 145-182 (1971)
2. Kauffman, S.A. *The origins of order.*, Oxford University Press (1993)
3. Kauffman, S.A. *At home in the universe.*, Oxford University Press (1995)
4. Serra, R., Villani, M., Semeria, A. *Genetic network models and statistical properties of gene expression data in knock-out experiments.*, J. Theor. Biol. 227, 149-157 (2004)
5. Serra, R., Villani, M., Graudenzi, A., Kauffman, S.A. *Why a simple model of genetic regulatory networks describes the distribution of avalanches in gene expression data.*, J. Theor. Biol. 246, 449-460 (2007)
6. Ramo, P., Kesseli, J., Yli-Harja, O. *Perturbation avalanches and criticality in gene regulatory networks.*, J. Theor. Biol. 242, 164-170 (2006)
7. Hughes T. R. et al., *Functional discovery via a compendium of expression profiles*, Cell, **102** (2000), 109–126.
8. Harvey, I., Bossomaier, T. *Time out of joint: Attractors in asynchronous random Boolean networks.*, in Husbands, P., Harvey, I., eds.: Proceedings of the Fourth European Conference on Artificial Life (ECAL97), 67-75, MIT Press (1997)
9. Aldana, M., Coppersmith, S., Kadanoff, L. *Boolean dynamics with random couplings.*, in Kaplan, E., Marsden, J.E., Sreenivasan, K.R., eds.: Perspectives and Problems in Nonlinear Science, Springer Applied Mathematical Sciences Series (2003)
10. Bastolla U. and Parisi G., *The modular structure of Kauffman networks*, Physica D, **115** (1998), 219–233.

CONTINUOUS NETWORK MODELS OF GENE EXPRESSION IN KNOCK-OUT EXPERIMENTS: A PRELIMINARY STUDY

A. ROLI and F. VERNOCCHI

DEIS, Campus of Cesena,
Alma Mater Studiorum *Università di Bologna*
Cesena, Italy
E-mail: andrea.roli@unibo.it, francesco.vernocchi@studio.unibo.it

R. SERRA

Dipartimento di scienze sociali, cognitive e quantitative,
Università di Modena e Reggio Emilia,
Reggio Emilia, Italy
E-mail: rserra@unimore.it

In this work we simulate gene knock-out experiments in networks in which variable domains are continuous and variables can vary continuously in time. This model is more realistic than other well-known switching networks such as Boolean Networks. We show that continuous networks can reproduce the results obtained by Random Boolean Networks (RBN). Nevertheless, they do not reproduce the whole range of activation values of actual experimental data. The reasons for this behavior very close to that of RBN could be found in the specific parameter setting chosen and lines for further investigation are discussed.

Keywords: Genetic networks, gene expression, knock-out gene, random boolean networks, Glass networks

1. Introduction

In previous studies,[1] it is shown that single gene knock-out experiments can be simulated in Random Boolean Networks (RBN), which are well-known simplified models of genetic networks.[2,3] The results of the simulations are compared with those of actual experiments in *S. cerevisiae*. The actual data are taken from the experiments described in a work by Hughes et al,[4] in which a genetic network of over 6000 genes is considered and a series of 227 experiments in which one gene is silenced are run on DNA microarrays.

Genes are knocked-out (i.e. silenced) one at a time, and the variations in the expression levels of the other genes, with respect to the unperturbed case (the *wild* type), are considered as the ratio of the activation in knock-out state (KO) and wild type (WT). Besides the ratios of KO/WT activation, *avalanches* can be defined that measure the size of the perturbation generated by knocking out a single gene.

In previous work on RBN,[1,5] it has been found that the distributions of avalanches are very robust, i.e. they are very similar in different random networks and the distribution of avalanches of the RBN models are close to those observed in actual experiments performed with *S. cerevisiae*. These results suggest that these distributions might be properties common to a wide range of genetic models and real genetic networks.

RBN are a very simplified model of genetic networks as they assume that a gene is either active or inactive, whilst in nature gene activation can range in a continuous domain. In this work we undertake an analogous study as in the case of RBN using a continuous network model, first proposed and discussed by Glass.[6,7] We show that Glass networks can reproduce the same results as RBN. Moreover, also the results of experiments with DNA microarrays are reproduced with a high level of accuracy. Nevertheless, with the parameter setting we used, this model is still not capable of capturing the whole range of KO/WT activation ratios.

Glass networks are described in Section 2 and their differences with Boolean Networks are outlined. Section 3 provides an overview of the main experimental settings and results are reported and discussed in Sections 4 and 5 in which we compare the results of Glass networks simulations with RBN and actual experiments, respectively. We conclude by discussing future work in Section 6.

2. Continuous networks

Glass networks[6,7] are continuous time networks in which node activation rate is regulated by a differential equation that includes a non-linear component that depends on a Boolean function of the node inputs. In these networks, time and gene activation are continuous, while the influence among genes is represented by switching functions. The activation of gene i is indicated with variable x_i ranging in a continuous domain. We associate to x_i a Boolean variable X_i defined as follows:

$$X_i(t) = \begin{cases} 0 \text{, if } x_i(t) < \theta_i \\ 1 \text{, otherwise} \end{cases}$$

In a network with N nodes, each with K inputs we define the activation rate of node i as:

$$\frac{dx_i}{dt} = -\tau_i x_i + f_i(X_{i_1}(t), X_{i_2}(t), \ldots, X_{i_K}(t)), \qquad i = 1, 2, \ldots, N$$

where f_i is a Boolean function of the inputs of node i.

Since the functions f_i change only when at least one variable crosses the threshold, the equations can be solved analytically in the intervals in which the Boolean functions are constant. Thus, if we denote with $T_s = \{t_1, t_2, \ldots, t_s\}$ the set of switching times, for each $x_i, i = 1, 2, \ldots, N$, and $t_j < t < t_{j+1}, t_j \in T_s$, we have:

$$x_i(t) = x_i(t_j)e^{-(t-t_j)\tau_i} + \frac{1}{\tau_i}f_i(X_{i_1}(t_j^*), X_{i_2}(t_j^*), \ldots, X_{i_K}(t_j^*))(1 - e^{-(t-t_j)\tau_i})$$

where $t_j^* \in (t_j, t_{j+1})$.

This model still introduces strong simplifications, but it explicitly takes into account the continuous time dynamics and continuous values of actual genetic networks. A more simplified model, though able to capture relevant properties of genetic networks, is that of (Random) Boolean Networks.[2] Variables associated to nodes in RBN assume values in the binary domain $\{0, 1\}$ and the transition functions are Boolean functions of the inputs. Usually, a synchronous dynamics is imposed to RBN. RBN have been studied as a model of genetic networks in which a gene is either active or inactive and it has been shown that they can simulate important properties of genetic networks.[2,5]

In this work, we present a preliminary study in which Glass networks are used to simulate gene knock-out experiments, as previously done with RBN. Our goal is to check if these networks can reproduce the results of simulations by RBN and to what extent and under which hypotheses they can capture more realistic features of the actual genetic networks.

3. Experimental setting

We performed a series of experiments to study the influence of single knock-out genes in continuous networks. Node activation ranging in a continuous domain makes it possible to compare results both with RBN, by converting values x_i into Boolean ones, and directly with *DNA microarray* results.

We designed and implemented a software system to simulate Glass networks and alike. The software has been designed trying to find a trade-off between performance, in terms of execution time, and extensibility. The tool can be configured so as to simulate models with different parameter settings and network characteristics, such as the topology. The simulator has been implemented in C++.

The networks we generated have a unitary decay parameter, $\tau_i = \tau = 1$, for every node and threshold value $\theta_i = \theta = 0.5$ $(i = 1, 2, \ldots, N)$, so as to have node values in the range $[0, 1]$. The other network parameters have been chosen according to previous work in which the results of simulations of RBN are compared with results in DNA microarrays. Thus, every node has two inputs, i.e., $K = 2$, randomly chosen among the other nodes. Boolean functions are assigned randomly to nodes, by picking them among the set of *canalizing* functions, i.e., functions in which at least one value of one of its inputs uniquely determines the output. For the case with $K = 2$, all the possible 16 functions except for *coimplication* and *exclusive or* are canalizing.

We generated 30 networks with 6000 nodes; each network is initialized with random values in the range $[0, 1]$ and its evolution in time is simulated until an attractor is reached.[a] The activation of a node is computed as the average value it assumes along the attractor. Then, 227 genes randomly chosen among the active ones, i.e., the ones with average activation greater than zero, were silenced in turn. The activation of the genes in the knock-out experiment were then evaluated in the new attractor and the ratio between the activation in the knock-out and wild type has been computed. Hence we obtain a 6000×227 matrix of real values that can be compared both with the corresponding matrix of experiments with RBN and real data. For each network we produce a matrix $E_{ij}, i = 1, \ldots, 6000$, $j = 1, \ldots, 227$, in which E_{ij} is the ratio of the activations of gene i in experiment j.

4. Comparison with Random Boolean Networks

The first analysis we make is a comparison of the results of simulations of gene knock-out in Glass networks with those of RBN. The values of the matrix built from real data and from simulation of the continuous model have been processed as in previous work[1] by introducing a threshold θ_E to define the level of meaningful difference between the knock-out and wild type: the difference between KO and WT activations is considered significant if the ratio is greater than θ_E or less than $1/\theta_E$. Hence we obtain a

---

[a]The simulation is stopped when, after a transient, a repetition of a network state is found; therefore, in these experiments, we classify the attractor found simply as a cycle.

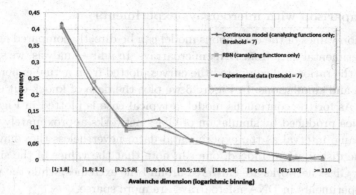

Fig. 1. Avalanche size frequency in simulations with continuous and Boolean network models and experiments in DNA microarray (linear binning).

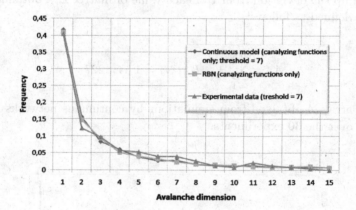

Fig. 2. Avalanche size frequency in simulations with continuous and Boolean network models and experiments in DNA microarray (logarithmic binning).

Boolean matrix E'_{ij} defined as follows: $E'_{ij} = 1$ if $E_{ij} > \theta_E \vee E_{ij} < 1/\theta_E$, $E'_{ij} = 0$, otherwise. As for results by RBN, any ratio not equal to 1 is considered as meaningful. The threshold θ_E has been set to 7, as from original work on RBN.[1]

Figures 1 and 2 plot the frequency of avalanche size of the two models and actual experimental data, in linear and logarithmic binning, respectively. This comparison shows that the results of the continuous model, when processed as the experimental ones, exhibit an avalanche frequency that closely approximates both the actual one and that of RBN.

5. Comparison with microarrays experiments

Activation values of the continuous model can be directly compared against the experimental data from DNA microarrays. In a first analysis, we simply ordered the ratios and compared the curves plotted from actual experimental data and simulations. In Figure 3 we plot the data of knock-out experiments. As for the continuous model, a typical case is plotted in Figure 4. The ratios produced by simulation of Glass networks approximately range in the same interval as the experimental data, nevertheless they have not the same distribution. Indeed, one can note that the values of the simulations by Glass networks are clustered around the extremes, while the values from experiments in DNA microarrays are more sparse.

We also considered a measure of the avalanche produced by a gene knock-out that does not depend upon a threshold. For each experiment we summed up the deviation from 1 of each value of matrix E_{ij}, obtaining an array A defined as follows:

$$A_j = \sum_{i=1}^{6000} |1 - E_{ij}| \ , \ j = 1, \ldots, 227$$

The array A obtained from simulations by continuous networks is the average over the 30 experiments.

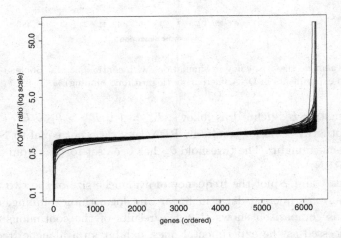

Fig. 3. KO/WT activation ratio in real data. In the x-axis, genes are ordered in non decreasing values of KO/WT activation ratio.

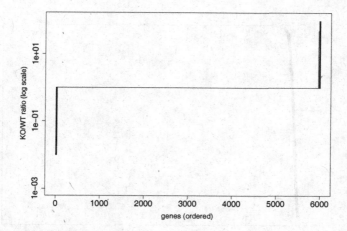

Fig. 4. KO/WT activation ratio in Glass networks. In the x-axis, genes are ordered in non decreasing values of KO/WT activation ratio.

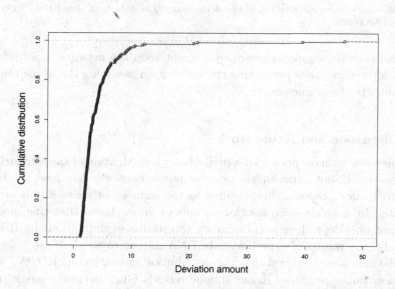

Fig. 5. Cumulative distribution of the deviations A_j in data from experiments.

In Figures 5 and 6 the cumulative distribution of the deviations is plotted in the case of experiments and simulations via continuous networks, respectively. The difference between the two distributions is apparent, as in the previous analysis. The discrepancy we observe could be ascribed to

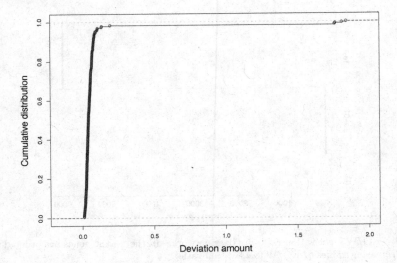

Fig. 6. Cumulative distribution of the deviations A_j in data from simulations by continuous networks.

the network parameters chosen, that might keep the network in a 'quasi-Boolean' regime, thus preventing the nodes from assuming the whole range of values in the attractors.

6. Discussion and future work

In this work we have presented a preliminary investigation of the simulation of gene knock-out experiments via continuous networks. We have studied the frequency of avalanches, defined as the number of genes significantly affected by a single gene knock-out, and we have shown that this model can not only reproduce with accuracy the results of simulations by RBN, but also the results of experiments in DNA microarrays.

We have also observed that the distribution of continuous KO/WT activation values produced in our simulations via Glass networks seems still not very close to that of actual experiments. However, this is a preliminary study and further analyses are planned in which crucial parameters of Glass networks will be varied in order to try a more accurate tuning of the model. First of all, delays τ_i and thresholds θ_i can be varied and be different across the nodes. In addition, different equations regulating the expression rate can be studied. Finally, the topology of the network can be changed and networks with more realistic topologies can be studied.

References

1. R. Serra, M. Villani and A. Semeria, *Journal of Theoretical Biology* , 149 (2004).
2. S. A. Kauffman, *The Origins of Order: Self-Organization and Selection in Evolution* (Oxford University Press, 1993).
3. S. Kauffman, *Current topics in dev. biol.* **6** (1971).
4. T. R. Hughes et al, *Cell* **102**, 109 (2000).
5. R. Serra, M. Villani, A. Graudenzi and S. A. Kauffman, *Journal of Theoretical Biology* , 449 (2007).
6. K. Kappler, R. Edwards and L. Glass, *Signal Processing* (2003).
7. R. Edwards, *Physica D* , 165 (2000).

SYNCHRONIZATION PHENOMENA IN INTERNAL REACTION MODELS OF PROTOCELLS

ROBERTO SERRA

Dipartimento di Scienze Sociali, Cognitive e Quantitative
Università di Modena e Reggio Emilia
via Allegri 9, 42100 Reggio Emilia, Italy
E-mail: rserra@unimore.it

TIMOTEO CARLETTI

Département de Mathématique, Facultés Universitaires Notre Dame de la Paix
8 rempart de la Vierge, B5000 Namur, Belgium
E-mail: timoteo.carletti@fundp.ac.be

ALESSANDRO FILISETTI

European Center for Living Technology
Ca'Minich, Calle del Clero, S.Marco 2940 - 30124 Venice, Italy
E-mail: alessandro.filisetti@ecltech.org

IRENE POLI

Dipartimento di statistica, Università Ca' Foscari
San Giobbe - Cannaregio 873, 30121 Venezia, Italy
E-mail: irenpoli@unive.it

1. Introduction

Protocells are lipid vesicles (or, less frequently) micelles which are endowed with some rudimentary metabolism, contain "genetic"material, and which should be able to grow, reproduce and evolve. While viable protocells do not yet exist, their study is important in order to understand possible scenarios for the origin of life, as well as for creating new "protolife"forms which are able to adapt and evolve[1] . This endeavor has an obvious theoretical interest, but it might also lead to an entirely new "living technology", definitely different from conventional biotechnology.

Theoretical models can be extremely useful to devise possible protocell architectures and to forecast their behavior. What can be called the "genetic material" of a protocell is composed by a set of molecules which, collectively, are able to replicate themselves. At the same time, the whole protocell undergoes a growth process (its metabolism) followed by a breakup into two daughter cells. This breakup is a physical phenomenon which is frequently observed in lipid vesicles, and it has nothing to do with life, although it superficially resembles the division of a cell. In order for evolution to be possible, some genetic molecules should affect the rate of duplication of the whole container, and some mechanisms have been proposed whereby this can be achieved.

In order to form an evolving protocells population it is necessary that the rhythms of the above mentioned two processes, i.e. methabolism and genetic replication, are synchronized and it has previously been shown that this may indeed happen when one takes into account successive generations of protocells[2-6] .

The present paper presents and extends our previous studies which had considered synchronization in the class of so–called "Internal Reaction Models", IRM for short[6] when linear kinetics were assumed for the relevant chemical reactions. Let us stress here that similar results have been obtained also for the "Surface Reaction Models", SRM for short[2-6] , hence the synchronization phenomenom seems to be very robust with the respect to the chosen architecture once linear kinetics are considered.

The IRMs are roughly inspired by the so–called RNA–cell[7,8] whereas the modelization of SRMs arises from the so-called "Los Alamos bug"[9,10] . The paper is organized as follows. In Sec. 2 we report a review of our previous results, in Sec. 3 we describe the main features of IRMs and discuss the behaviors of this class of models. Finally, in Sec. 4 some critical comments and indications for further work are reported.

2. A review of previous results

As already explained in our previous works[2-6] starting from a set of simplified hypotheses and considering a protocell endowed with only one kind of genetic memory molecule, where the relevant reactions occurs on the external protocell surface one can describe the container and genetic molecule behavior by Eqs. (1).

$$\begin{cases} \frac{dC}{dt} = \alpha C^{\beta-\gamma} X^{\gamma} \\ \frac{dX}{dt} = \eta C^{\beta-\nu} X^{\nu}, \end{cases} \tag{1}$$

where C is the total quantity of the "container" material, β is a parameter that determines the thickness of the container (ranging between 2/3 for a micelle and 1 for a very thin vesicle), X is the total quantity of the genetic memory molecule and γ and ν are positive parameters related to the rates of the chemical reactions.

Before going on with the discussion about the internal reactions models it is interesting to consider which kind of behaviors one can expect to find:

(1) Synchronization: in successive generations (as $k \to \infty$ where k is the generation number) the interval of time needed to duplicate the membrane molecules of the protocell between two consecutive divisions, ΔT_k, and the time required to duplicate the genetic material, again between two consecutive divisions, ΔT_k^g, approach the same value;

(2) as $k \to \infty$ the concentration of the genetic material at the beginning of each division vanishes. In this case, given the above assumptions, the growth of the container ends and the whole process stops;

(3) as $k \to \infty$ the concentration of the genetic material at the beginning of each division cycle, grows unbounded. This points to a limitation of the equations introduced before, that indeed lack a rate limiting term for the growth rate of X;

(4) the two intervals of time, ΔT_k and ΔT_k^g, oscillate in time (we will provide some examples in the following) with the same frequency . This condition is not equivalent to synchronization *strictu sensu* but it would nonetheless allow sustainable growth of the population of protocells. Therefore this condition might be called *supersynchronization*. Note that in principle supersynchronization does not require equality of the two frequencies, but that their ratio be a rational number;

(5) the two intervals of times, ΔT_k and ΔT_k^g, change in time in a "chaotic way".

3. Internal reaction models

Let us now consider the synchronization problem in a model where the relevant chemical reactions are supposed to run inside the protocell vesicle, such models have been named internal reaction models (IRMs)[6] .

Assuming once again that the genetic memory molecules induce the container growth via the production of lipids from precursors, one can describe the amount of container C in time by some non–linear function of the concentration $[X] = X/V$:

$$\frac{dC}{dt} = \alpha X^\gamma V^{1-\gamma}, \tag{2}$$

where $\gamma > 0$ is a parameter which determines the strength of the influence of X on C and where V is the whole protocell volume, V, namely we are here assuming that the volume occupied by the amount of C is really small with respect to the volume determined by X

Let us now consider some possible replication rates for the genetic molecules. The simplest one is the *Linear replicator kinetics*; in this case the amount of the X molecules is proportional to the number of existing ones (given that precursors are not limiting), so:

$$\frac{dX}{dt} = \eta X. \tag{3}$$

A straightforwardly generalization can be obtained assuming some power law with exponent ν, of the concentration $[X]$, hence we get:

$$\frac{dX}{dt} = \eta [X]^\nu V = \eta X^\nu V^{1-\nu}. \tag{4}$$

The behavior of a protocell during the continuous growth phase is thus described by:

$$\begin{cases} \frac{dC}{dt} = \alpha X^\gamma V^{1-\gamma} \\ \frac{dX}{dt} = \eta [X]^\nu V = \eta X^\nu V^{1-\nu}. \end{cases} \tag{5}$$

In order to complete the treatment it is necessary to express V as a function of C, and this depends upon geometry. One can assume a generic functional relation of the form $V = g(C)$, for some positive and monotone increasing function g.

To provide an explicit example we will now compute such function g under the assumption of spherical vesicle with very thin membrane.

Remark 3.1 (Spherical very thin vesicle). *Let us suppose that the vesicle is spherical, with internal radius r_i and with a membrane of constant width δ (a reasonable assumption if it is a bilayer of amphiphilic molecules). Then starting from $V = V_i + V_C$ we get:*

$$V_i = \tfrac{4}{3}\pi r_i^3 \text{ and } V_C = \tfrac{4}{3}\pi(t_i + \delta)^3 - \tfrac{4}{3}\pi r_i^3 \Rightarrow V_C = 4\pi r_i \delta^2 + 4\pi r_i^2 \delta + \tfrac{4}{3}\pi \delta^3. \tag{6}$$

We can thus express r_i as a function of C and then using the formula for the sphere volume we can express V as a function of C, through its dependence on r_i. *One can easily obtain:*

$$r_i = \frac{-\delta^2 + \sqrt{-\frac{1}{3}\delta^4 + \frac{V_C}{4\pi}}}{2\delta}, \tag{7}$$

we can finally assume $\delta << 1$ (thin membrane), to get

$$V_C \cong 4\pi r_i^2 \delta = S\delta \tag{8}$$

where S is the surface area, $S = 4\pi r_i^2$, and finally assuming $V_C << 1$,

$$V \sim V_i = \frac{4}{3}\pi r_i^3 = \frac{4}{3}\pi \left(\frac{S}{4\pi}\right)^{\frac{3}{2}} \cong \frac{4}{3}\pi \left(\frac{V_C}{4\pi\delta}\right)^{\frac{3}{2}} = \frac{4}{3}\pi \left(\frac{C}{4\pi\delta\rho}\right)^{\frac{3}{2}}. \tag{9}$$

Thus the required function is $V = aC^{3/2}$.

By incorporating the constants into the kinetic constants and renaming them, the model (5) can thus be described by

$$\begin{cases} \frac{dC}{dt} = \alpha X^\gamma C^{3(1-\gamma)/2} \\ \frac{dX}{dt} = \eta X^\nu C^{3(1-\nu)/2}. \end{cases} \tag{10}$$

The model described by Eq. 5, or by 10, can be studied via an analytical technique presented in[2] and[6]. Here we propose an alternative approach that enables us to obtain the same results and also some explanations.

The division event can be seen as a map that to the amount of the X–molecule at the beginning of the k–th generation arising at time T_k, associates the same quantity, say X_{k+1} at the beginning of the next protocell cycle:

$$F : (X_k, T_k) \mapsto F(X_k, T_k) = (X_{k+1}, T_{k+1}). \tag{11}$$

Then synchronization is *equivalent* to determine a *fixed point* for this map, if moreover we are interested in the possibility to reach this fixed point following the dynamics, this fixed point must be a *stable* one.

The map F can be obtained by integrating Eqs. (5). To simplify the successive computations we first introduce an analytical trick, consisting in a non–linear reparametrization of time. In fact from the first relation of Eq. (5) we can conclude that C is a monotone increasing function of time, i.e. its derivative is strictly positive, hence we can introduce a new time variable, τ, defined by:

$$\frac{d\tau}{dt} = \alpha \left[g(C)\right]^{1-\gamma} X^\gamma dt, \tag{12}$$

where $V = g(C)$ denotes the generic dependence of the volume on the container C. Using this new variable the system (5) can be rewritten as:

$$\begin{cases} \frac{dC}{d\tau} = 1 \\ \frac{dX}{d\tau} = \frac{\eta}{\alpha} X^{\nu-\gamma} g^{\gamma-\nu} . \end{cases} \qquad (13)$$

In this way the behavior of C is trivial and the division event is just $\tau_{k+1} = \tau_k + \theta/2$. This simplifies the map F that becomes a function of X_k only. Moreover during the continuous growth we have $C(\tau) = \theta/2 + (\tau - \tau_k)$ and thus the second relation of (13) rewrites:

$$\frac{dX}{d\tau} = \frac{\eta}{\alpha} X^{\nu-\gamma} \left[g\left(\theta/2 + (\tau - \tau_k)\right)\right]^{\gamma-\nu} , \qquad (14)$$

This equation can be solved explicitly thus providing the map F:

$$F(X_k) = \left[\frac{1}{2^{\gamma-\nu+1}} X_k^{\gamma-\nu+1} + (\gamma - \nu + 1)\frac{\eta}{\alpha} \int_0^{\theta/2} g(\theta/2 + s)\, ds \right]^{1/(\gamma-\nu+1)}$$

$$\text{if } \gamma - \nu + 1 \neq 0. \qquad (15)$$

The case $\gamma - \nu + 1 = 0$ can be solved as well but one can show that in this case synchronization is possible only for special values of the involved parameters and thus it is not generic. For this reason we will not develop further this case.

The function F admits a positive fixed point if and only if $\gamma - \nu + 1 > 0$ which is given by:

$$F(X_*) = X_* \Rightarrow X_* = \left(\frac{\gamma - \nu + 1}{2^{\gamma-\nu+1} - 1} \frac{\eta}{\alpha} \Phi(\theta) \right)^{1/(\gamma-\nu+1)} , \qquad (16)$$

where we denoted by $\Phi(\theta)$ the constant integral in the right hand side of (15).

To determine the stability character of this fixed point we have to compute the first derivate of F and evaluate it at X_*, that is:

$$\frac{dF}{dX}(X_*) = \frac{X_*^{1/(\gamma-\nu+1)}/2^{\gamma-\nu+1}}{\Phi(\theta) + X_*^{1/(\gamma-\nu+1)}/2^{\gamma-\nu+1}} , \qquad (17)$$

and we can easily check that under the assumption $\gamma - \nu + 1 > 0$ and $X_* > 0$ this derivative is always smaller than 1, ensuring thus the stability of the fixed point.

This result can thus be restated by saying that synchronization is possible only if the genetic replication rate is small enough with respect to the container growth: $\nu < \gamma + 1$.

Remark 3.2. The very widely used model of quadratic growth for genetic memory molecules doesn't allow for synchronization if the container growth is linear, in fact here the previous relation doesn't hold, i.e. $2 = \nu \nleq \gamma+1 = 2$. But observe that if the container growth were slightly faster, say $\gamma > 1$, then synchronization will be obtained.

$$\begin{cases} \frac{dC}{dt} = \alpha X \\ \frac{dX}{dt} = \eta \frac{X^2}{V} . \end{cases} \tag{18}$$

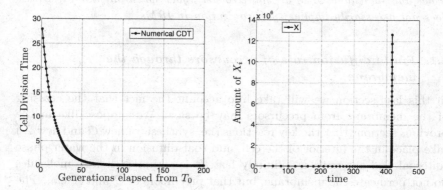

Fig. 1. An example of a system ruled by Eqs. (18) where synchronization is not achieved. On the left panel cell division time in function of generations elapsed from $T_{(0)}$ is shown while on the right panel the total amount of replicators in function of time for each generation is shown.

Remark 3.3 (Various kinetic equations). *In the case where more than one kind of genetic memory molecules is present in the same protocell, one can consider of course more general situations as already done for the SRM case.*[2,4] *Under the assumption of very thin spherical vesicle we have considered the following two cases of second order interaction between different genetic memory molecules:*

(1) ***without cross-catalysis:*** *no synchronization is observed for the studied set of parameters:*

$$\begin{cases} \frac{C}{dt} = \alpha_1 X_1 + \cdots + \alpha_n X_n \\ \frac{dX_i}{dt} = C^{-\frac{3}{2}} \sum_{k=1}^{N} M_{ik} X_i X_k \end{cases} \tag{19}$$

(2) with cross-catalysis :

$$\begin{cases} \frac{C}{dt} &= \alpha_1 X_1 + \cdots + \alpha_n X_n \\ \frac{dX_i}{dt} &= C^{-\frac{3}{2}} \sum_{k=1}^{N} M_{ijk} X_j X_k . \end{cases} \tag{20}$$

The behaviour is not completely understood. Varying the kinetic coefficients sometimes we observe synchronization but more often extinction.

Similar results hold true for the SRM case as well, but on different time scales and somehaow SRM are more robust infact synchronization in SRMs does not necessarily implies synchronization in IRMs.

3.1. *Finite diffusion rate of precursors through the membrane*

In this last section we will take into account the fact that the crossing of the membrane from precursors may be slow. We suppose like in the previous sections that the key reactions (i.e. synthesis of new C and new X) take place in the interior of the cell, and that diffusion in the water phase (internal and external) is infinitely fast. It is assumed that X molecules do not permeate the membrane, but that precursors of C and X can. The external concentration of these precursors is buffered to fixed values E_C and E_X, while the internal concentrations can vary, their values being $[P_C] = P_C/V$ and $[P_X] = P_X/V$, where V is the inner volume, thus once again we assume that the membrane volume to be negligeable. Note that, for convenience, the fixed external concentrations are indicated without square brackets, while P_C and P_X denote internal quantities.

Precursors can cross the membrane at a finite rate; if D denotes diffusion coefficient per unit membrane area, then the inward flow of precursors of C (quantites/time) is $D_C S(E_C - [P_C])$, and a similar rule holds for X.

X catalyzes the formation of molecules of C, therefore we assume that the rate of growth of C is proportional to the number of collisions of X molecules with C precursors in the interior of the vesicle. It is therefore a second order reaction. Reasoning as it was done in the case of Sec. 3 one gets

$$\begin{cases} \frac{dC}{dt} &= \alpha' h_C V^{-1} X P_C \\ \frac{dX}{dt} &= \eta' h_X V_i^{-1} X P_X . \end{cases} \tag{21}$$

Note that it might happen that more molecules of precursors are used to synthesize one molecule of product (the number of precursor molecules per product molecule can be called h_X and h_C).

$$\begin{cases} \frac{dP_X}{dt} = SD_X \left(E_X - \frac{P_X}{V}\right) - \eta' h_X V^{-1} X P_X \\ \frac{dP_C}{dt} = SD_C \left(E_C - \frac{P_C}{V}\right) - \alpha' h_C V^{-1} X P_C, \end{cases} \tag{22}$$

Equations (21) and (22) provide a complete description of the dynamics. Note that by defining $\eta = \eta' h_X$ and $\alpha = \alpha' h_C$ one can eliminate the stoichiometric coefficients from these equations.

As done before, in order to complete the study it is necessary to express V and the surface S as a functions of C, and this depends upon geometry. Under the assumption of spherical very thin vesicle we obtain

$$V = aC^{\frac{3}{2}} \quad \text{and} \quad S = bC, \tag{23}$$

for some positive constants a and b. The second relation of Eq. (23), inserted in Eq. (22), complete the model. The behavior of this model has been studied with numerical methods, Fig. 2, and it has been numerically verified that this model shows synchronization in the range of considered parameters.

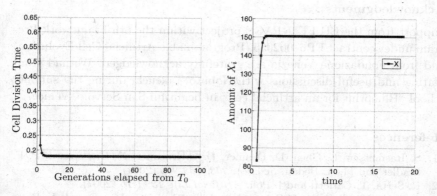

Fig. 2. An example of a system ruled by Eqs. (21 and (22) where synchronization is achieved. On the left panel cell division time in function of generations elapsed from $T_{(0)}$ is shown while on the right panel the total amount of replicators in function of time for each generation is shown.

4. Conclusion

In this paper we have addressed some relevant questions about synchronization in a class of abstract models of protocell called Internal Reaction Models (IRMs) where key reactions occur within the vesicle, this complete our previous works where all reactions occurred on the surface of the protocell (SRMs).

Comparing the two classes of models we observe that the behavior is very similar in the two cases so synchronization is an emergent property also of IRMs.

We also demonstrated that synchronization is an emergent property independently from the geometry of the container if the genetic replication rate is small enough respect to the container growth, Sec. 3.1.

Most of the analyses have been carried on under the simplifying assumption that diffusion through the membrane is fast with respect to the kinetic constants of the other processes. Since this may be unrealistic in some real cases, we have also considered a case with finite diffusion rate, showing the way in which such a case can be modelled and demonstrating, under the particular kinetic model considered, that synchronization is also achieved.

It is worth remarking that, although the properties which have been shown in this paper provide a clear picture of synchronization, further studies are needed in order to consider more general cases.

Acknowledgments

Support from the EU FET–PACE project within the 6th Framework Program under contract FP6–002035 (Programmable Artificial Cell Evolution) and from Fondazione Venezia are gratefully acknowledged. We had stimulating and useful discussions with Nobuto Takeuchi during the summer school "Blueprint for an artificial cell" in beautiful San Servolo, Venice.

References

1. S. Rasmussen, L. Chen, D. Deamer, D. C.Krakauer, N. H. Packard, P. F. Stadler and M. A. Bedau, *Science, 303, 963-965* (2004).
2. R. Serra, T. Carletti and I. Poli, *Artificial Life 13: 1-16* (2007).
3. R. Serra, T. Carletti, I. Poli and A. Filisetti, *In G. Minati and A. Pessa (eds): Towards a general theory of emergence. Singapore: World Scientific (in press)* (2007).
4. R. Serra, T. Carletti, I. Poli, M. Villani and A. Filisetti, *Accepted to ECCS-07: European Conference on Complex Systems* (2007).

5. R. Serra, T. Carletti and I. Poli, *BIOMAT2006, International Symposium on Mathematical and Computational Biology, (World Scientific, ISBN 978-981-270-768-0)* (2007).
6. T. Carletti, R. Serra, I. Poli, M. Villani and A. Filisetti, *submitted* (2008).
7. T. O. et al., *Biochemical and Biophysical Research Communications 207: 250-257* (1995).
8. D. Szostak, P. Bartel and P. Luisi, *Nature 409: 387-390* (2001).
9. S. Rasmussen, L. Chen, M. Nilsson and S. Abe, *Artificial Life, 9, 269-316* (2003).
10. S. Rasmussen, L. Chen and B. M. Stadler, *Origins Life and Evolution of the biosphere, 34, 171-180* (2004).

SEMI-SYNTHETIC MINIMAL CELLS[*]

PASQUALE STANO

Biology Department, University of RomaTre, Viale G. Marconi 446
Rome, 00146, Italy

PIER LUIGI LUISI[†]

Biology Department, University of RomaTre, Viale G. Marconi 446
Rome, 00146, Italy

In this short article we would like to review some of our recent achievements concerning the laboratory construction of minimal cells. Minimal cells are cell-like molecular systems containing the minimal number of components in order to be defined as "alive". According to the current approach, since these minimal cells are built from synthetic compartments (liposomes) and extant genes and enzymes, they have been dubbed as semi-synthetic ones, and are now considered within one of the key projects of synthetic biology. Minimal cells, on the other hand, thanks to their minimal functional processes, are often referred as model of early cells, and may provide insights into the nature of primitive cells.

1. Introduction

1.1. *Minimal living cells*

The question "what is life?" and the idea of constructing simple forms of life in the laboratory, have been in the agenda of science for a very long time. A first assessments to the question "what is life", as well as the approach to the construction of simple life forms in the laboratories, have to start from the following basic consideration: no other form of life on earth other than the cellular life. This means that life is based on compartments that permit a high local concentration of reagents, protection from the environment, containment and control of structural changes. These compartments are defined by semi-

[*] This work has been supported by the "Enrico Fermi" Study and Research Centre, Rome, Italy. It is currently supported by the EU Synthcells project (Contract Nr. 043359) and by the Human Frontiers Science Program.

[†] Corresponding Author. Phone and Fax (+39) 06 5733 6329; e-mail: luisi@mat.ethz.ch

permeable membranes that permit the selection of chemicals as nutrients or other co-adjuvant of the cell functions.

The genealogy of the minimal cells idea, from the biological, chemical and philosophical approaches, has been recently reviewed in several works [1-5]. Aspects as the continuity principle, minimal complexity, definition of life, minimal life, requirements for being alive, and others, bring us to conceive a research line that focuses on the construction and study of minimal artificial constructs that are designed and realized with the aim of achieving a set of minimal living functions.

The first thing that is necessary for implementing in the laboratory the construction of a minimal cell is a suitable compartment. Lipid vesicles (liposomes) represent a proper model for cellular membrane. The strategy consists in the insertion of the minimal number of genes and enzymes inside a synthetic vesicle, so as to have an operational viable cell. The fact that natural macromolecules are used for this kind of construct, brings to the terminology of "semi-synthetic minimal cells", meaning that part of this construct is synthetic (the membrane, and the assemblage processes), whereas some other parts (enzymes and nucleic acids) are natural. With these premises, the next point is to clarify what do we mean by "a minimal number of genes and enzymes". The question of minimal genome is strictly connected to this issue. In order to be considered alive, a cellular construct must satisfy some conditions, which have been indicated in several reviews [1-5]. These theoretical considerations find a realistic scenario in the studies of several researchers, who point out what is the minimal number of genes in a living organism. Thanks to the available data on minimal genomes it is possible to evaluate a minimal gene-set of 200-300 genes. The interested reader can refer to a recently published review [2]. Here we would like to comment one of the most recent study, a contribution of A. Moya and coworkers, who, on the basis of a comparative and functional genomics approach, indicates 206 genes as the genomic core required for minimal living forms. The analysis was carried out by a comparative approach that considers the genomes of five endosymbionts and other microorganisms [6]. Such small number of genes codifies for the proteins that perform the essential cell functions, as basic metabolism and self-reproduction. A pictorial representation of the Moya's minimal genome is shown in Figure 1. To date, investigations on verifying the minimal genome number, and on further reduction of it, are necessarily only speculative. Minimal cells, however, will help in solving this question experimentally, so that a minimal genome could be established experimentally on the basis of future experiments.

Our current work, in contrary, focuses on experimental approaches to the construction of minimal cells.

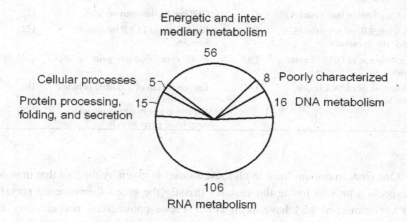

Figure 1. The functional classification of a minimal genome composed of 206 genes. Data from reference [6].

2. Literature stand

A survey of literature data indicates that the large majority of the researchers in this field have approached the construction of minimal semi-synthetic cells by performing simple and complex compartmentalized biochemical reactions into lipid vesicles. In the last years, several papers appeared in the literature, describing the insertion of complex biochemical systems in vesicles. Several groups, in particular those lead by T. Yomo [7-9], T. Ueda [10-12], and T. Sugawara [13,14] in Japan, our group in Italy [15-23], D. Deamer [24,25], S. Rasmussen [26], as well as A. Libchaber [27,28] in the United States, E. Szathmáry in Hungary [29], Y. Nakatani and the late G. Ourisson in France [30,31] are currently involved in projects that deals with minimal cells. In particular it has been possible to express some proteins by inserting the whole ribosomal synthetic apparatus into vesicles. The most representative results have reported in Table 1.

Table 1. Compartmentalized reactions in lipid vesicles.

Nucleic Acid Metabolism	Ref.	Protein Expression	Ref.
Enzymatic poly(A) synthesis inside vesicles	[32,33]	Ribosomal synthesis of polypeptides into vesicles	[19]
RNA replication into vesicles	[18]	GFP production into vesicles	[7]
DNA amplification by the PCR inside the liposomes	[17]	Expression of EGFP into small vesicles	[22]
Transcription of DNA in giant vesicles	[30]	EGFP expression into giant vesicles	[31]
Production of mRNA inside giant vesicles	[34]	Encapsulation of a genetic network into vesicles	[8]
		Expression of two proteins into a long-lived giant vesicle	[28]

One first, important limit in all these studies is given by the fact that in order to express a protein inside the vesicles (mostly the green fluorescence protein, GFP), commercial kits have been used. Those commercial preparations are usually "black boxes", in the sense that the composition of the mixture is not given, and therefore also the number and the relative concentration of the enzymes involved are unknown. Since all these preparations are cellular extracts (from *E. coli*), a rigorous synthetic approach is not possible.

Another limit, clearly arising from the analysis of literature data, is that only proteins have been expressed, which corresponds to an active internal metabolism; however there is no result pointing to self-reproduction of these semi-synthetic cells. Our work stems from the effort of overcoming these two important limits of present day literature. In particular, we wish to carry out the research by employing a minimal, and well-known, set of enzymes, and focusing in the self-reproduction.

3. Recent achievements in the construction of the semi-synthetic minimal cells

3.1. *Protein expression in vesicles by means of the "PURESYSTEM®"*

Few years ago, the group of Takuya Ueda at the Tokyo University reported on the creation of a new in vitro protein expression kit composed by purified components [10-12]. This kit, now commercially available with the trademark of PURESYSTEM® (Post Genome Institute Co., Ltd. – Japan), is composed by 36 purified enzymes, t-RNAs and ribosomal components; each one of these compounds is present at a known concentration. Originally developed to synthesize proteins in vitro, this tool appears to be perfectly suitable for a pure synthetic biology approach to the minimal cell projects. In fact, it allows the construction of an artificial cell that contains a minimal number of components in order to perform some function, i.e., the synthesis of a functional protein – in this case.

We have introduced for the first time the use of PURESYSTEM in microcompartmentalized reactions, in order to avoid the use of cellular extracts (from Promega, Roche, Ambion, etc.). The advantages deriving from utilizing PURESYSTEM instead of these "black boxes" are manifold:

a) the use of a macromolecular mixture of known composition (molecular species, number and amount of each);

b) the possibility to tune the performance of the mixture by changing the composition of the mixture;

c) the possibility to remove further components in order to reduce the number of components required for specific functions, e.g. expression of a functional protein, etc.;

d) the establishment of a standard approach for the construction of semi-synthetic minimal cells.

In order to express a functional protein inside liposome, the PURESYSTEM was encapsulated in liposomes together with the pWM/T7/EGFP plasmid, which encodes the gene of the enhanced green fluorescent protein (EGFP) a useful probe to quickly verify the synthesis of functional proteins inside vesicles. Here, to prevent the EGFP synthesis at the outside of liposomes, the RNase A was added to the reaction mixture soon after liposome formation. To measure the EGFP fluorescence produced within 1-palmitoyl-2-oleoyl-*sn*-glycero-3-phosphatidylcholine (POPC) liposomes, we have used fluorometric analysis. Figure 2 shows the EGFP fluorescence intensity (508 nm), as expected for the green fluorescence of this molecule (after correction for the Rayleigh scattering). The total amount of synthesized EGFP inside liposomes was estimated as almost 10% of the yield in bulk.

Confocal microscopy was used in order to obtain further evidence about the localization of protein synthesis inside the liposome (data not shown) [35].

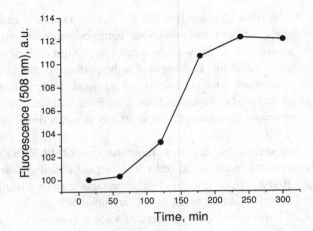

Figure 2. Fluorometric analysis of EGFP produced within liposome. From ref. [35]

3.2. *Lipid-synthesizing liposomes: the salvage pathway*

Self-reproduction in the case of semi-synthetic cells is a difficult problem that should be addressed in various degrees of approximation. The first level that we propose is that of a system that expresses in its core those enzymes, which catalyze the synthesis of the membrane (Figure 3). Preliminary studies have been carried out some years ago in our group, aimed at producing inside lecithin liposomes the synthesis of lecithin [15]. According to these early results and also to some new developments [36,37], the synthetic route that appears to be promising is the so-called lipid salvage pathway, indicated in Figure 3.

The advantage of this pathway is also related to the chemical nature of precursors. In fact, in addition to glycerol-3-phosphate, the long-chain acyl-CoA, needed to carry out steps 1 and 2 of the route, as well as CDP-choline in step 4 are all water-soluble compound. In other words, the salvage pathway transforms small water-soluble molecules into a membrane-forming lipid molecule, i.e., lecithin.

As additional possibility, we must consider that in principle it is possible to stop the salvage pathway after the second acylation step, obtaining phosphatidic acid, a molecule that forms per se lipid bilayers and vesicles. We recently investigated the formation and the properties of dioleoylphosphatidic acid vesicles, and its interaction with oleoyl-CoA [38].

Therefore, the practical approach to such study involves the entrapment of the PURESYSTEM kit together with the four genes codifying for the four enzymes of lipid biosynthesis, e.g., the expression of lipid-synthesis enzyme-battery (Figure 3a).

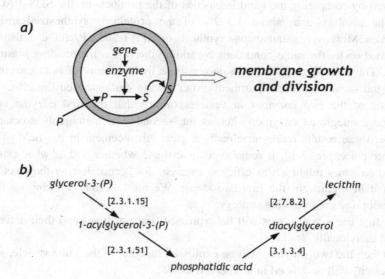

Figure 3. A liposome that enclosed a metabolic network, which in turn leads to the production of the membrane-forming compounds "S" (a). Lecithin biosynthesis by the salvage pathway (b). The four enzymes needed to accomplish the transformation of glycerol-3-phosphate into phospatidylcholine are: *sn*-glycerol-3-phosphate *O*-acyl-transferase [2.3.1.15], 1-acyl-*sn*-glycerol-3-phosphate *O*-acyl-transferase [2.3.1.51]; phosphatidate phosphatase [3.1.3.4]; diacylglycerol cholinephosphotransferase [2.7.8.2]. The reagents needed in each step are omitted.

We proceeded then to study the synthesis of two enzymes involved in the biosynthesis of phosphatidic acid in *E. coli*: the *sn*-glycerol-3-phosphate acyltransferase (GPAT) and the 1-acyl-*sn*-glycerol-3-phosphate acyltransferase (LPAAT). Both proteins are localized in the cytoplasmic membrane. The molecular size of GPAT is estimated as 83 kDa and the LPAAT as 27.5 kDa. The aim of synthesizing these proteins in liposome is to produce endogenously phospholipids by means of internalized enzymes. The synthesis of these proteins is demonstrated by their radiolabeling (using radio-labeled [^{35}S] methionine in the system). The synthesis of GPAT and LPAAT is successfully observed at expected position on the SDS–PAGE gel. When this reaction mix is not encapsulated within POPC liposomes, protein synthesis was completely inhibited by RNase A. On the contrary, when the same reaction is encapsulated in liposomes, protein synthesis is observed even in the presence of external RNase A. The ratio of the inside synthesis versus the whole synthesis can be evaluated by comparing the band intensities of the products on the SDS–PAGE gel. The result is that about 3.5–9% of the protein is synthesized inside liposomes. Moreover, simultaneous synthesis of GPAT and LPAAT can also be performed under the same conditions by adding the two corresponding plasmid DNAs. The next target will be to demonstrate the synthesis of both enzymes within the same vesicle compartment. After having demonstrated the effective synthesis of the two enzymes in vesicles (notice that the first enzyme is a membrane integrated enzyme, whereas the second is a membrane-associated enzyme; these results represent already a great advancement in the field of in vitro protein expression), it remains to investigate whether and at what extent the two enzymes might act as efficient catalysts for accomplishing the reaction indicated in Figure 3b (the first two steps). We are currently working on this topic, adopting the following strategy:

a) first the two enzymes will be expressed (once at time) and their activity studied individually

b) then the two enzymes will be expressed together in the same vesicle, and their activity will be probed in a "coupled assay".

To date, GPAT and LPAAT show an intermediate and a strong activity, respectively. The reasons that could affect the low efficiency of GPAT are still unclear. They can be related to a minor amount of expressed enzyme due to intrinsic low productivity, or to misfolding, or to other effects, as product inhibition, or poor assay conditions, or problems related to the permeability of substrates. The work is in progress – to the aim of coupling the two enzymes that should work as an enzyme battery, and to transform the substrates into phosphatidic acid.

4. Concluding remarks

The research work done in the last years represents an innovative approach to understand the nature of cellular life and some aspects of the origins of life by means of constructive studies. Although this research area is still at its infancy, the interest raised by such investigation suggests that the significance of this topic is going to increase in the next years, also because its connection to synthetic biology. From a more general viewpoint, it is important to emphasize that semi-synthetic minimal cells represent a concrete approach to 'artificial life' based on real molecules and processes. With the creation of more sophisticated models, it might be experimentally shown that life is indeed an emergent property originating when a critical ensemble of macromolecules gives rise to self-organized structures and processes that produce themselves, as indicated by the autopoietic theory [39].

References

1. P. L. Luisi *The Emergence of Life*. Cambridge University Press, Cambridge (2006).
2. P. L. Luisi, F. Ferri and P. Stano, *Naturwissenschaften* **93**, 1 (2006).
3. P. Stano, F. Ferri, and P. L. Luisi, in: *"Life as we know it"*, ed. J. Seckbach, Springer, Berlin, pp. 181-198 (2006)
4. P. L. Luisi, P. Stano, G. Murtas, Y. Kuruma and T. Ueda T, in: *"Proceedings of the Bordeaux Spring School on Modelling Complex Biological Systems in the Context of Genomics. April 3rd - 7th 2006"*. P. Amar; F. Képès; V. Norris; M. Beurton-Aimar; J.-P. Mazat (eds.). EDP Sciences, Les Ulis (Paris), pp. 19-30 (2007).
5. P. Stano, G. Murtas and P. L. Luisi, in: *"Protocells: Bridging Nonliving and Living Matter"*, MIT Press. S. Rasmussen and M. Bedau (eds.), in press.
6. R. Gil, F. J. Silva, J. Peretó and A. Moya, *Microbiol. Mol. Biol. Rev.* **68**, 518 (2004).
7. W. Yu, K. Sato, M. Wakabayashi, T. Nakatshi, E. P. Ko-Mitamura, Y. Shima, I. Urabe and T. Yomo, *J. Biosci. Bioeng.* **92**, 590 (2001).
8. K. Ishikawa, K. Sato, Y. Shima, I. Urabe and T. Yomo, *FEBS Letters* **576**, 387 (2004).
9. T. Sunami, K. Sato, T. Matsuura, K. Tsukada, I. Urabe and T. Yomo, *Analyt. Biochem.* **357**, 128 (2006).
10. Y. Shimizu, A. Inoue, Y. Tomari, T. Suzuki, T. Yokogawa, K. Nishikawa and T. Ueda, *Nature Biotech.* **19**, 751 (2001).
11. Y. Kuruma, K. Nishiyama, Y. Shimizu, M. Mueller and T. Ueda, *Biotech. Progr.* **21**, 1243 (2005).
12. Y.Shimizu, T. Kanamori and T. Ueda, *Methods* 36, 299 (2005).
13. K. Takakura and T. Sugawara, *Langmuir* **20**, 3832 (2004).
14. K.Takakura, T. Toyota and T. Sugawara, *J. Am. Chem. Soc.* **125**, 8134 (2003).
15. P. K. Schmidli, P. Schurtenberger and P. L. Luisi, *J. Am. Chem. Soc.* **113**, 8127 (1991).
16. P. Walde, A. Goto, P. A. Monnard, M. Wessicken and P. L. Luisi, *J. Am. Chem. Soc.* **116**, 7541 (1994).
17. T. Oberholzer, M. Albrizio and P. L. Luisi, *Chem. Biol.* **2**, 677 (1995).
18. T. Oberholzer, R. Wick, P. L. Luisi and C. K. Biebricher, *Biochem. Biophys. Res. Commun.* **207**, 250 (1995).

19. T. Oberholzer, K. H. Nierhaus and P. L. Luisi, *Bioch. Biophys. Res. Commun.* **261**, 238 (1999).
20. J. W. Szostak, D. P. Bartel and P. L. Luisi, *Nature* **409**, 387 (2001).
21. P. L. Luisi, *Anatom. Rec.* **268**, 208 (2002).
22. T. Oberholzer and P. L. Luisi, *J. Biol. Phys.* **28**, 733 (2002).
23. S. Islas, A. Becerra, P. L. Luisi and A. Lazcano, *Orig. Life Evol. Biosph.* **34**, 243 (2004)
24. A. Pohorille and D. Deamer, *Trends Biotech.* **20**, 123 (2002).
25. D. Deamer, *Trends Biotech.* **23**, 336 (2005).
26. S. Rasmussen, L. Chen, D. Deamer, D. C. Krakauer, N. H. Packard, P. F. Stadler and M. A. Bedau, *Science* **303**, 963 (2004).
27. V. Noireaux and A. Libchaber, *Proc. Natl. Acad. Sci. USA* **101**, 17669 (2004)
28. V. Noireaux, R. Bar-Ziv, J. Godefroy, H. Salman and A. Libchaber, *Phys. Biol.* **2**, P1 (2005).
29. E. Szathmary, M. Santos and C. Fernando, *Top. Curr. Chem.* **259**, 167 (2005).
30. K. Tsumoto, S. M. Nomura, Y. Nakatani and K. Yoshikawa, *Langmuir* **17**, 7225 (2001).
31. S. M. Nomura, K. Tsumoto, T. Hamada, K. Akiyoshi, Y. Nakatani and K. Yoshikawa, *ChemBioChem* **4**, 1172 (2003).
32. A. C. Chakrabarti, R. R. Breaker, G. F. Joyce and D. W. Deamer, *J. Mol. Evol.* **39**, 555 (1994).
33. P. Walde, A. Goto, P. A. Monnard, M. Wessicken and P. L. Luisi, *J. Am. Chem. Soc.* **116**, 7541 (1994).
34. A. Fischer, A. Franco and T. Oberholzer, *ChemBioChem* **3**, 409 (2002).
35. G. Murtas, Y. Kuruma, P. Bianchini, A. Diaspro and P. L. Luisi, *Biochem. Biophys. Res. Comm.* **363**, 12 (2007).
36. P. Luci, *ETH-Z Dissertation* No. 15108 (2003).
37. Y. Kuruma, *Orig. Life Evol. Biosph.* **37**, 409 (2007).
38. J. Dubois, P. Stano and P. L. Luisi, *unpublished data* (2005).
39. P. L. Luisi, *Naturwissenschaften* **90**, 49 (2003).